European Agency for Safety and Health at Work

EUROPEAN RISK OBSERVATORY REPORT

Expert forecast on Emerging Biological Risks related to Occupational Safety and Health

European Agency
for Safety and Health
at Work

Authors:
European Agency for Safety and Health at Work:
Emmanuelle Brun

Topic Centre Risk Observatory:
Steve Van Herpe, Prevent, Belgium
Irja Laamanen, FIOH, Finland
Colette Le Bâcle, INRS, France
Kerstin Klug, Gunter Linsel and Rüdiger Schöneich, BAuA, Germany
Eva Flaspöler & Dietmar Reinert, BGIA, Germany
Magdalena Galwas, CIOP, Polen
Maria Asunción Mirón Hernández and Daniel García-Matarredona Cepeda, INSHT, Spain

In cooperation with:
Topic Centre Risk Observatory:
Jean-Marie Mur, INRS, France
Ellen Zwink, BAuA, Germany
Eulalia Carreras, INSHT, Spain

Cover photos:
1. Courtesy of Istituto Zooprofilattico Sperimentale delle Venezie, Italy.
2. Farming.
3. Courtesy of European Commission.

Cataloguing data can be found at the end of this publication.

Luxembourg: Office for Official Publications of the European Communities, 2007

ISBN 92-9191-130-5

Printed in Belgium

PRINTED ON WHITE CHLORINE-FREE PAPER

TABLE OF CONTENTS

Foreword

Working environments are continuously changing with the introduction of new technologies, substances and work processes, changes in the structure of the workforce and the labour market, and new forms of employment and work organisation. New work situations bring new risks and challenges for workers and employers, which in turn demand political, administrative, technical and regulatory approaches to ensure high levels of safety and health at work.

In 2000, the Lisbon summit identified specific objectives to create quality jobs and increase workforce participation. Improving working conditions to keep people in work is necessary if these objectives are to be achieved. In this context, the Community strategy on health and safety at work 2002–06 called on the European Agency for Safety and Health at Work (the Agency) to 'set up a risk observatory'. One of its priorities would be to 'anticipate new and emerging risks, whether they be linked to technical innovation or caused by social change'. The strategy emphasized that this should be done by 'ongoing observation of the risks themselves, based on the systematic collection of information and scientific opinions', as part of the development of a 'genuine culture of risk prevention'.

The Agency, therefore, took the first step towards establishing a European Risk Observatory, commissioning its Topic Centre Risk Observatory (TCRO) — the former Topic Centre Research on Work and Health (TCWH) — which includes some of the principal OSH institutions in Europe, to identify emerging risks related to OSH. To this end, two types of activities have been carried out: the collection of published information from reliable sources — still ongoing — and the production of expert forecasts.

The expert forecasts on emerging OSH risks were reached through questionnaire-based surveys following the Delphi method. Four Delphi surveys have been carried out: on physical risks; psychosocial risks; chemical risks; and biological risks. This division into four themes was neither meant to indicate fixed boundaries between the areas nor to exclude combinations of them. On the contrary: many OSH issues are multifactorial and have been mentioned in several of the surveys. In total, 520 experts from 27 countries and one international organisation were invited to participate in the surveys. Answers were received from 188 experts from 24 countries and one international organisation, giving a response rate of 35%.

This report is the second of a series of European Risk Observatory reports dedicated to emerging risks. It sets out an expert forecast on emerging biological OSH risks. The results of this forecast have also been used as a basis for discussion among representatives from major European OSH research institutes and from UNICE, ILO, DG Research and DG Employment in a seminar organised by the Agency aimed at promoting occupational safety and health research in the EU (Bilbao, Spain, 1st and 2nd December 2005). Several of the emerging issues identified in the forecast have been included in a summary list of top OSH research priorities drawn up at the seminar and consolidated in a broader consultation process among the Agency's stakeholders. Using this list, the OSH research community can present a clear message during the seventh framework programme (FP7) consultation to promote the inclusion of OSH issues.

The Agency would like to thank the members of the Topic Centre Risk Observatory for their contributions to the drafting of this report. Most of all, it would like to thank all the safety and health experts from around Europe who took time to reply to the survey; their participation was essential to the project. The Agency would also like to thank its focal points, Expert Group and Advisory Group for their valuable comments and suggestions.

European Agency for Safety and Health at Work

Executive summary

Context

This report contains a forecast of emerging biological risks related to occupational safety and health (OSH) based on an expert survey and a literature review. The Agency also worked on forecasts and literature reviews on physical, chemical, and psychosocial risks in order to paint as full a picture as possible of the potential emerging risks in the world of work.

These results are linked to other Risk Observatory work and aims to examine OSH trends in Europe and to anticipate emerging risks and their likely consequences for safety and health at work. This should help with better targeting of resources and lead to more timely and effective interventions.

Method

Within the scope of this project, an 'emerging OSH risk' has been defined as any occupational risk that is both new and increasing.

By new, it is meant that:

- the risk was previously non-existent; or
- a long-standing issue is now considered a risk due either to new scientific knowledge, or to a change in social or public perceptions.

The risk is increasing if:

- the number of hazards leading to the risk is growing; or
- the likelihood of exposure to the hazard leading to the risk is increasing; or
- the effect of the hazard on workers' health is getting worse.

For the formulation of the expert forecast on emerging OSH biological risks, a questionnaire-based survey was run in three consecutive rounds following the Delphi method. This method was chosen to avoid individual, non-scientifically founded opinions, and to verify whether a consensus could be reached among the respondents. Some 109 experts in the first survey round and 95 experts from each of the second and third rounds were invited to participate in the survey following their nomination by the Agency's focal points and Topic Centre Research. Thirty-two valid questionnaires from the first round, 42 from the second and 36 from the third were returned from 58 organisations in 18 Member States, as well as Bulgaria, Romania and Switzerland. The response rates were 29% (first round), 44% (second) and 38% (third). Participating experts were required to have at least five years' experience in the field of OSH and biological risks. Respondents were mainly involved in research, consulting or teaching and training activities, followed by labour inspection and policy development.

Respondents were required to have at least five years' experience in the field.

The 'top' emerging biological risks identified

Occupational risks related to global epidemics are the biggest emerging issue identified in this forecast, with a high level of consensus among the respondents. Even in the 21st century, we are still confronted with the emergence of new pathogens, such as severe acute respiratory syndrome (SARS) or avian influenza, and the re-emergence

When a new contagious pathogen emerges, its worldwide spread is inevitable and is likely to be very rapid.

of outbreak-prone diseases such as cholera and yellow fever. When a pathogen emerges, given the speed and volume of international traffic and trade, it may spread around the world within a few hours and start a new pandemic.

Cases of SARS in health care staff, the 89 Dutch poultry workers infected with the avian virus A/H7N7 in 2003, workers unloading international trade containers facing risks of contamination with Dengue fever from mosquitoes imported with the goods, the current threat of avian flu A/H5N1 to poultry workers and culling workers are only a few examples that show how epidemics can affect the world of work.

More than three quarters of human diseases are zoonoses.

As more than three quarters of these diseases are zoonoses, workers at risk are those who are in:

- contact with live or dead infected animals
- contact with aerosols, dust or surfaces contaminated by animal secretions
- in animal trade, breeding and slaughtering facilities,
- cleaning and disinfection jobs in contaminated areas
- veterinary services, research laboratories, customs, zoos and pet shops.

Workers involved in global trade, air crews and air travellers, those working to control epidemics, some media professionals, and workers in war zones, such as those offering peacekeeping or humanitarian aid assistance, are also at risk. Additionally, healthcare staff, who may be exposed to infected individuals, are a high-risk group.

Multi-disciplinary cooperation is needed to tackle the global threat of pandemics.

Practical guidance on how to protect workers from risks related to global epidemics is already available for some diseases. As infected workers may in turn spread the disease among the general public, OSH considerations have to be urgently integrated into public health pandemic plans. More generally, cooperation between various authorities, including public health, occupational health, animal health, food safety, and environmental protection, is of the utmost importance. Several risk factors are known to increase the chances of outbreaks. These include microbiological adaptation, globalisation of transport, trade, agriculture and food production, human behavioural factors and environmental changes. Although it is difficult to predict future outbreaks, the systematic monitoring of these factors is essential to the effective forecasting, surveillance, prevention and control of future epidemics and pandemics.

Drug-resistant organisms pose a risk to workers in contact with animals and in the health care sector.

A further emerging risk, which also illustrates the importance of cooperation between various disciplines and authorities, is the **emergence of drug-resistant organisms**. Since their discovery in the 20th century, antimicrobial agents have substantially reduced the threat of infectious diseases. However, this advantage is now jeopardised by the emergence and worldwide spread of antimicrobial-resistant organisms, mainly as a result of the overuse or misuse of antibiotics. Resistant organisms pose a health risk to workers in contact with animals — for example, in veterinary services and in the livestock and food-manufacturing industry — and to healthcare workers in hospitals with the emergence of organisms such as methicillin resistant staphylococcus aureus (MRSA). Drug-resistant organisms lead to severe infections that would not otherwise occur, and to more failures in treatment. Measures to stop the spread of such organisms and the contamination of workers, include the improvement of work organisation, regular cleaning of the work premises, use of safety-engineered sharp-instruments, appropriate handling of clinical waste and thorough hand washing. Further recommendations overlap with the public health sphere and advocate the strict control of antibiotics use.

Along these lines, an EU-wide ban on the use of antibiotics as growth promoters in animal feed came into effect on January 1st 2006. Still, it is inevitable that antimicrobial-resistant organisms will continually evolve. The challenge is to identify them quickly as they emerge, assess their potential impact on health, identify the sources and routes of exposure, and devise policies and procedures to minimise their spread. While there is an urgent need for harmonisation of experimental conditions between the various studies being undertaken on antimicrobial resistance, international conventional surveillance systems need to be adapted to local situations because different habits of antibiotic usage result in different spectrum and susceptibility patterns of invasive pathogens in the different Member States.

Validated exposure assessment methods and dose-effect relationships are needed to assess properly biological risks.

Risks resulting from **poor risk assessment** are the second most important of the emerging issues. Directive 2000/54/EC lays down the principles for the management of biological risks and assigns to employers the duty of assessing the risks posed by biological agents in the workplace. But the state of knowledge on biohazards is still relatively immature and, in practice, proper assessment of biological risks is difficult. In order to produce a proper exposure assessment, better tools for the detection of biological agents and measurement of their concentrations need to be developed. These should be based on non-culture techniques because culture methods have proven to be of limited use. The validation of measurement methods and international harmonisation of those methods are also necessary if laboratory results are to be comparable. Such harmonisation should include the definition of commonly approved criteria and accepted protocols for assessing exposure to biological hazardous substances; including concise and uniform guidelines on sampling, storage, extraction and analytical procedures. This, together with more epidemiological and clinical data, is the basis for understanding better the relationships between exposure and occupational health effects. Of course, the actual effect depends on an individual's susceptibility. Information on dose-effect relationships would also help to establish occupational exposure limits (OELs), which, conversely, would support the proper interpretation of measurement results in a risk assessment procedure. As at October 2006, although some Member States have formulated recommendations and set indicative values, very few obligatory OELs have been set for airborne microorganisms or their associated toxins.

The **lack of information on biological risks** in the workplace, which makes risk assessment difficult, has been treated as an emerging risk in a separate item, especially in the office workplace and the agriculture sector. Furthermore, the lack of information passed on to workers — i.e. the **inadequate provision of OSH training to workers**, especially in local authorities — has also been raised.

A further emerging risk, with a rather high consensus among the experts, is the **poor maintenance of water and air systems.** This puts workers — and the general population — at risk of legionella. Moreover, this again illustrates the consequences of having insufficient information on biohazards. Indeed, the experts comment that some ill-health symptoms observed in indoor workers are often wrongly assumed to be flu-like diseases. In fact, they are caused by biological agents that have developed in poorly maintained air-conditioning systems. Recent findings on legionella will help establish a correct diagnosis of such symptoms.

More research into combined exposure to biological agents and chemicals is needed.

If the risks engendered by biological agents are difficult to assess, **combined exposure to biological agents and chemicals** is all the more challenging, and is actually closer to the reality of workplaces. While the range of potential subsequent health effects is wide, it is difficult to determine which of the constituents primarily accounts for which health effects. In addition, more research is needed to help identify the real multi-factorial causes of health symptoms, for which mono-causal explanations are often incorrectly given.

The gaps and challenges in terms of exposure assessment, establishment of dose-effect relationships and OELs, as well as the risk assessment mentioned earlier, are particularly true for endotoxins and mould, which have also been singled out as emerging risks in this report.

Endotoxins are found in organic dust.

Endotoxins can be found in high concentrations in all occupational settings where organic dust is present. What was initially considered to be a problem in only a few activities turned out to also affect workers in the livestock industry, scientists working with rodents, workers in waste and sewage treatment and even indoor workers. Research in recent years has revealed major clinical effects of endotoxins, ranging from fever, infectious diseases, acute toxic effects, allergies, organic dust toxic syndrome (ODTS), chronic bronchitis, and asthma-like syndromes, to lethal effects such as septic shock, organ failure and death. However, a complex relationship between exposure to endotoxins and the outcome of immune responses has been found. Indeed, although seemingly paradoxical, endotoxins may induce, but conversely also protect from, asthma, atopy, respiratory allergies and sensitisation to allergens. The protection and prevention measures recommended include: minimising the generation of organic dust; moving work activities outdoors whenever possible or operating within a controlled atmosphere if indoors; and as a final protection measure, wearing appropriate personal protective equipment.

Moulds may be present in any indoor workplace.

Indoor moulds and subsequent health issues have only been given greater attention relatively recently. To date, more than 100,000 species of mould have been identified, but it is estimated that there may be more than 1.5 million species worldwide. As airborne moulds are ubiquitous in the indoor environment, workers in any indoor workplace, such as offices, schools, hospitals, homes and commercial buildings, may be exposed. Airborne mould is also found in waste and sewage treatment activities, in cotton mills and in the agricultural sector. Additionally, hazardous materials removal workers and construction workers involved in the remediation of mould-contaminated areas, which is a new and growing part of their work, are at risk. Health-based exposure limits to airborne mould could not yet be established. The most common symptoms induced by mould are sick building syndrome (SBS), asthma, upper respiratory diseases, infections, coughs, headaches and flu-like symptoms, allergic diseases, and irritation of the nose, throat, eye and skin. Although it is not possible to completely eliminate mould spores, it is possible to control moisture, which is one of the factors promoting mould growth. For this purpose, the 'health' of a building should be addressed before building starts, with cooperation between all the people involved in the building construction, design, use and maintenance. In the case of fungal contamination of a building, prompt remediation of contaminated material must be undertaken. Recommendations for the safe handling, disposal, recycling, and transportation of mouldy materials are available. However, there is still a need for reliable criteria and measurement methods so it can be determined with certainty that the remediation has been successful.

More and more workers are exposed to biological agents in the waste treatment industry.

The occupational risks linked to **waste treatment** have been identified as emerging in this forecast, as well as in a similar survey on chemical risks ([1]). In the 1990s, several governments and the European Union adopted new waste management policies with the primary aim of decreasing the amount of waste sent to landfill. The recycling industry is a relatively new but expanding business, employing a steadily growing number of workers. As the waste treatment regulation was primarily developed for

([1]) European Agency for Safety and Health at Work, 'Expert forecast on emerging chemical risks related to occupational safety and health'. The report will be published in 2007.

environmental purposes, it insufficiently addresses OSH issues. The major health problems observed in workers, especially in composting activities, are caused by bioaerosols. These include upper airway inflammations and pulmonary diseases, ODTS, gastrointestinal problems, allergic reactions, skin diseases, and irritation of the eyes and mucous membranes. Such bioaerosols mainly result from the generation of organic dust and contain a diversity of airborne microorganisms, including mould, as well as their toxic products such as endotoxins and volatile organic compounds (VOCs). Handling medical waste and sharps may lead to other infections, such as hepatitis and human immunodeficiency virus (HIV). Prevention should be adapted to each specific waste branch and activity. While it is not possible to eliminate completely the risks posed by biological agents in waste-related activities, the most efficient prevention measure is to reduce the generation of dust and aerosols. Collective prevention measures and hygiene plans can also greatly reduce workers' exposure.

1.

INTRODUCTION

In the last decade, media coverage has raised public awareness of biological hazards, such as the introduction of anthrax into occupational settings due to bioterrorist activities, the outbreak of severe acute respiratory syndrome (SARS) affecting health care workers and, more recently, the threat of avian flu to workers who have had contact with poultry. Biological agents are ubiquitous and, in many workplaces, workers are faced with less publicised but still considerably harmful biological hazards. As an example, indoor moisture is found even in relatively new buildings and causes serious health problems such as asthma and allergies in construction, maintenance and office workers. In many countries, blood-borne occupational infection with hepatitis C virus is a major epidemiological problem among health care workers. Hazardous bioaerosols are associated with a wide range of health effects and are present in many jobs, from agricultural and waste treatment activities to the metallurgy industry, where gram-negative bacteria are contained in aerosols arising from metalworking fluids. Biological agents — defined in Directive 2000/54/EC (2) as bacteria, viruses, fungi, cell cultures and human endoparasites able to provoke any infection, allergy or toxicity — are sometimes introduced intentionally into a work process; for instance in a microbiology laboratory or in the food industry, or may be undesirable but inherent to the job, such as in farming or waste treatment activities.

5,000 workers die each year of communicable diseases in the EU. In France, 15% of the workforce is exposed to biological agents.

Globally, an estimated 320,000 workers worldwide die every year of communicable diseases caused by viral, bacterial, insect and animal related biological hazards (3). Although most of these take place in developing countries, some 5,000 fatalities will occur in the European Union. Women are more likely to be hit than men as they work more in occupations that involve biological hazards and exposure (4).

Viruses, bacteria or parasites are responsible for at least 15% of all new cases of cancer worldwide (5). In 2001, around 1,900 cases of recognised occupational diseases in the EU-15 were due to biological agents (6). In France, 2.6 million workers were exposed to biological agents in their jobs in 2003, which represents 15% of the workforce (7). More than half of those were employed in health and social work, where two thirds were in contact with biological agents. Significant exposure to biological agents was also found in agriculture, the manufacture of food products, services to individuals and households, research and development, and sanitation activities.

(2) Directive 2000/54/EC of the European Parliament and of the Council of the 18th September 2000 on the protection of workers from risks related to exposure to biological agents at work (seventh individual directive within the meaning of Article 16(1) of Directive 89/391/EEC). Official Journal L 262. http://europa.eu.int/eur-lex/lex/LexUriServ/LexUriServ.do?uri=CELEX:32000L0054:EN:HTML

(3) Driscoll, T., Takala, J., Steenland, K., Corvalan, C., Fingerhut, M., 'Review of estimates of the global burden of injury and illness due to occupational exposures', *American Journal of Industrial Medicine,* 2005, World Health Organization, http://www.who.int/quantifying_ehimpacts/global/3gbdcomparison.pdf

(4) Takala, J., 'Introductory Report: Decent Work — Safe Work', *XVIth World Congress on Safety and Health at Work (Vienna, 27 May 2002)*, International Labour Office (ILO), SafeWork, http://www.ilo.org/public/english/protection/safework/wdcongrs/ilo_rep.pdf

(5) Bosch, F. X. et al. 'Infections', *UICC Handbook for Europe*, International Union Against Cancer, 2004, http://www.uicc.org/fileadmin/manual/9.6infections.pdf

(6) Karjalainen, A., Niederlaender, E., 'Occupational diseases in Europe in 2001', *Statistics in focus*, 15/2004. European Communities, 2004, ISSN: 1024-4352, http://epp.eurostat.ec.europa.eu/cache/ITY_OFFPUB/KS-NK-04-015/EN/KS-NK-04-015-EN.PDF

(7) Guignon, N., Sandret, N., 'Les expositions aux agents biologiques dans le milieu du travail en 2003', DARES, Premières Informations et Premières Synthèses, No 26.1, June 2006. http://www. Travail.gouv.fr/IMG/pdf/2006.06-26.1.pdf

Directive 2000/54/EC lays the principles for the management and prevention of biological risks. The key to minimising the risks posed to workers by biological agents lies in their proper assessment, as described in the Directive, which sets out the obligations of employers 'to determine and assess the risks in any activity where workers may be exposed to biological agents. Practical guidance on how to do so, as well as preventive tools, can be found on the Agency website dedicated to European Week 2003 (8) 'Dangerous substances, handle with care', which aimed at raising awareness and promoting activities to reduce the risks of working with dangerous substances, including biological agents. Other valuable information on biological agents and hazards in the workplace, such as the latest research findings and good practices, are continuously being gathered and made available in the emerging risks (9) and good practice (10) web sections of the Agency.

This report sets out to present the results of the Agency's work on biological agents in the workplace, an expert forecast on emerging biological OSH risks. The risks identified in this forecast have been thematically grouped into four categories:
- substance-specific biological risks
- biological risks intrinsic to specific workplaces or work processes
- biological risks resulting from risk management and prevention practices, and
- occupational biological risks linked to social and environmental phenomena.

Six literature reviews explore in more depth some of the main emerging risks singled out in the forecast in terms of context, workers at risk, health and safety outcomes, and prevention.

What are emerging risks?

An 'emerging OSH risk' has been defined as any occupational risk that is both new and increasing.

By new, it is meant that:
- the risk was previously unknown and is caused by new processes, new technologies, new types of workplace, or social or organisational change; or
- a long-standing issue is newly considered as a risk due to a change in social or public perceptions; or
- new scientific knowledge allows a long-standing issue to be identified as a risk.

The risk is increasing if:
- the number of hazards leading to the risk is growing; or
- the likelihood of exposure to the hazard leading to the risk is increasing (exposure level and/ or the number of people exposed); or
- the effect of the hazard on workers' health is getting worse (seriousness of health effects and/ or the number of people affected).

(8) http://ew2003.osha.eu.int/

(9) http://riskobservatory.osha.eu.int

(10) http://riskobservatory.osha.eu.int/

Three further forecasts have been produced on emerging physical [11], chemical [12], and psychosocial risks [13], in order to provide as comprehensive a picture as possible of the world of work. These activities are part of the European Risk Observatory, which aims to highlight OSH trends in Europe, provide an early identification of newly emerging risks in the workplace, and identify areas and issues where more information is needed.

(11) European Agency for Safety and Health at Work, 'Expert forecast on emerging physical risks related to occupational safety and health'. Belgium 2005. ISBN: 92-9191-165-8. http://riskobservatory.osha.eu.int/risks/forecasts/physical_risks/full_publication_en.pdf

(12) European Agency for Safety and Health at Work, 'Expert forecast on emerging chemical risks related to occupational safety and health'. To be published in 2007.

(13) European Agency for Safety and Health at Work, 'Expert forecast on emerging psychosocial risks related to occupational safety and health'. To be published in 2007.

2.

METHODOLOGY

2.1. Implementation of the expert survey

European experts were surveyed for their knowledge on the emerging OSH biological risks. The Delphi method was used in order to reach a broad consensus and to avoid non-scientifically founded opinions.

Delphi method ([14])

The Delphi method is a widely used methodology to build up information on topics for which only uncertain or incomplete knowledge is available. There are several variations of the Delphi method, but all of them are based on an iteration process with at least two survey rounds in which the results of the previous rounds are fed back and submitted again to the experts for new evaluation. The feedback process ensures that the experts are aware of the views of other experts and gives them the possibility to revise their first evaluation. At the same time it limits the chances of individuals being unduly influenced by group pressures, which could lead to experts not daring to offer their real opinions and lead to distorted results.

The Delphi method adopted for formulating an expert forecast on emerging risks in this project consisted of three survey rounds (see Figure 1). Only the answers from experts eligible for participation were analysed (see '3.1.Selection of participants').

Figure 1. Delphi process implemented for the expert forecast on emerging OSH biological risks

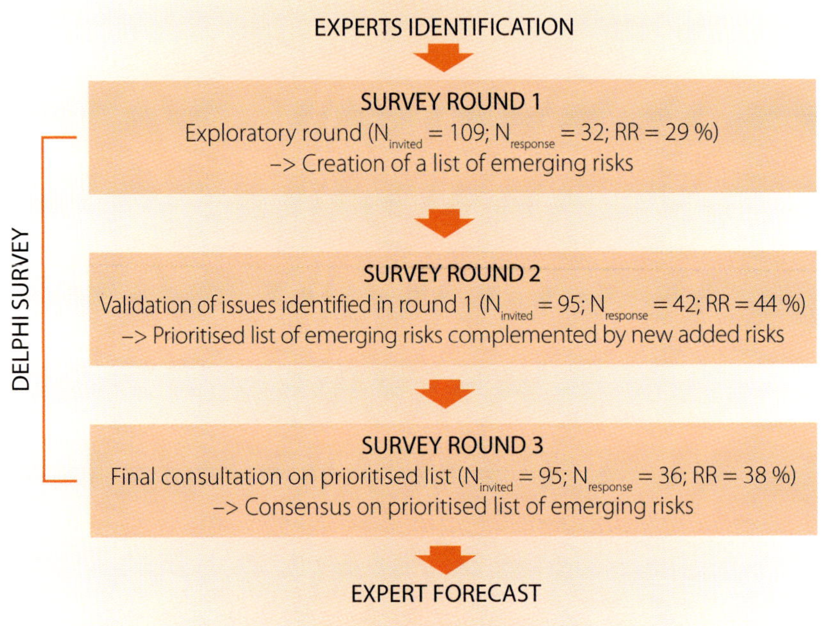

([14]) Cuhls, K., 'Technikvorausschau in Japan — Ein Rückblick auf 30 Jahre Delphi-Expertenbefragungen', *Technik, Wirtschaft und Politik*, Vol. 29. Schriftenreihe des Fraunhofer-Instituts für Systemtechnik und Innovationsforschung (ISI), Physica, Heidelberg, 1998

First survey round

A first exploratory survey round carried out in 2004 aimed to identify the emerging risks. A questionnaire with open-ended questions was developed to help the experts in formulating their views on the emerging OSH biological risks of the next 10 years. The experts were invited to fill in the questionnaire, either electronically or on paper. Based on all the issues identified in the returned questionnaires, a list was drawn up in which the risks could be sorted into four categories: substance-specific biological risks; biological risks specific to workplaces or resulting from specific work processes (risks related to recycling and waste handling activities, to the health care and service sectors, to laboratory and research activities, to the food industry, and to the agricultural sector were mentioned); risks resulting from risk management practices and handling; and OSH biological risks linked to social and environmental phenomena.

Second survey round

A second questionnaire-based survey round was carried out in 2005. This aimed to validate and complement the results of the first round. The questionnaire presented a list, drafted out of the first round, with feedback on the number of times each item was suggested. It invited participants to rate each item, independently from the others, on a five-point Likert scale (non-comparative scaling process). The scale ranged from 'strongly disagree that the issue is an emerging risk', through 'undecided' to 'strongly agree that the issue is an emerging risk'. The experts could add new risks to the list.

As a result of the second survey round, a prioritised list of risks was drawn up based on the mean values of the item ratings and the standard deviations (see box below for more details).

Third survey round

As the last step towards reaching a consensus, a third consolidation round was carried out in 2005.

The third questionnaire also consisted of a non-comparative scaling process whereby the respondents were asked to rate each issue independently from the others on the same five-point Likert scale used in the second round.

How to interprete the ratings

For each risk, the mean values and the standard deviations were calculated. While the mean values help to prioritise the risks within one risk category, the standard deviations reflect the level of consensus on one item among the respondents.

The following areas have been defined for the interpretation of the mean values, based on the definition of the five-point Likert scale used in the survey (see above), and in order to have a reasonable balance of items between the different areas:

- the risk is strongly agreed to be emerging if the mean value of the rating is above four ($MV > 4$);
- a mean value between 3.25 and 4 means that the item is considered to be an emerging risk ($3.25 < MV \leq 4$);

- as a mean value is unlikely to be exactly equal to 3, the 'undecided' area has been extended from 2.75 to 3.25, which means that the status of a risk is regarded as undecided if its mean value is within this interval (2.75 ≤ MV ≤ 3.25);
- there is agreement that the risk is not emerging if the mean value is between 2 and 2.75 (2 ≤ MV < 2.75);
- there is strong agreement that the risk is not emerging if the mean value is below 2 (MV < 2).

The prioritised lists of emerging risks established at the end of the third survey round form the expert forecast on emerging OSH biological risks.

2.2. Reliability of the data

For each item, the response data sets were checked for statistical anomalies (ratings deviating significantly from the median of the data). No specific respondent profile could be associated to the few exceptional ratings found. As the anomalies had no significant influence on the mean value, they were not removed from the data sets.

Kolmogorov-Smirnow-tests were also run in order to verify the standard distribution of the data.

Delphi studies usually end after two to four survey rounds ([15]). With regards to the present Delphi survey, a consensus among participants was reached in the third round for the majority of the items. Indeed, when considering only the answers from the 24 experts who responded to *both* the second *and* the third survey rounds, a comparison of the standard deviations (SDs) of round two with round three shows that most SDs decreased from one round to the next: out of the 36 items rated in both rounds, 19 SDs decreased and one did not vary. Considering this positive development, and also with a view to the limited financial resources and time allocated to the project, it was decided to end the Delphi survey at the third round.

Although the same experts were invited to participate in the second and third rounds, different persons actually responded to one or other round. In order to decide whether to base the forecast *only* on the answers of the participants in round three who *also* responded to round two (N=24), or on *all* answers from all participants to the third round (N=36) — including those who did not participate in the second round — the mean values were calculated for both population samples separately and

([15]) Fraunhofer-Institut für Systemtechnik und Innovationsforschung (ISI), 'Delphi 98 — Studie. Befragung zur globalen Entwicklung von Wissenschaft und Technik. Zusammenfassung der Ergebnisse.' http://www.isi.fraunhofer.de/p/Downloads/Delphi98-Methoden.pdf

compared. Globally, the mean values do not vary significantly between the two groups. With regards to the top ten items (Diagram 1), the differences lie between 0.04 points (for 'poor maintenance of air-conditioning and water systems' and 'inadequate OSH training') and 0.19 points (for 'biohazards in waste treatment plants'). Therefore, the ratings from *all* the experts who participated in the third survey round have been taken into consideration in order to have a forecast based on more participants.

Diagram 1. Comparison between the mean values of the 'top' ten items of the 3rd survey round for the following two population samples: all respondents to the third survey round (N=36); and respondents to both round 2 and round 3 (N=24)

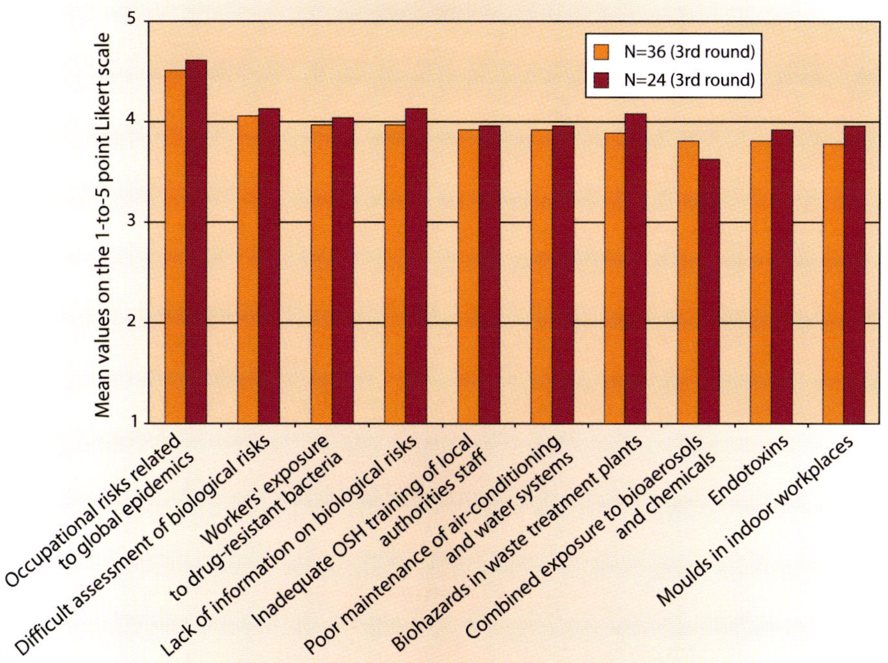

2.3. Limitations of the methodology

The study relies on the goodwill of the experts to complete the questionnaires, with no financial compensation for their contribution. What is more, the respondents had to understand written English and to be able to formulate their answers in English as the questionnaires were not translated. This has certainly had an effect on the response rate and may be one of the report's major limitations. Indeed, the higher the number of participants, the better the reliability and representativity of the forecast. In addition, it means that some countries may be over-represented — as was the case for Germany in this survey — depending on the country of origin of the experts willing to participate. This may affect the representativeness of the forecast in terms of the European view.

There are also limitations with the initial phase in which risks are defined. Analysing and compiling the free-text answers to the open-ended questions raised in the first, 'brainstorming' survey round is a difficult exercise. Indeed, the answers received were variable in terms of the amount of information and details provided, the level of specificity of the issues brought up — while some issues mentioned were, for example, substance-oriented, others had been formulated with a view to health outcomes, or overlapped with several other items but addressed only one specific workplace, or one specific health outcome — and the quality of the written English. As opposed to a workshop, in such a questionnaire-based process there is no opportunity for a moderator to ask the participants for clarification, to re-focus their answers on OSH when they have moved beyond the scope of the study, or, conversely, to provide them with the information they may need to answer the question adequately. These factors impede the setting of clear risk descriptions, which is essential to avoid misunderstanding on the items to be rated in the further rounds.

A further issue is the difficulty of finding the right participants. On the one hand, respondents with a deep but specific expertise may be too focused on their own area of work and mention only their own topics and activities in the survey. Conversely, generalists with broader knowledge may lack the expertise to judge whether an issue is actually emerging, and may be influenced by more political views.

Furthermore, the items identified in the survey are long-standing issues requiring action, rather than new or potential risks. Even if something has been known for some time, it can still be an emerging risk, because the scientific knowledge that enables us to understand that something is a risk is often deferred. Still, it seems that the point of the emerging risks definition referring to new risks — 'the risk was previously unknown and is caused by new processes, new technologies, new types of workplace, or social or organisational change' — has been poorly addressed in this exercise. Questionnaire-based surveys may not be suitable for the forecast and anticipation of issues that are genuinely new or do not yet exist.

Last but not least, because of the nature of forecasting activities, the evidence may still be inconclusive for some of the emerging risks mentioned in the survey. However, this does not mean that such issues should be avoided; this would mean that the Risk Observatory had failed to accomplish its main objective. Rather, particular care should be taken to discuss the findings with the relevant stakeholders in order to validate any conclusions and decide on the need for any further work on the topic. In this way, the Risk Observatory will fulfil its mission to stimulate debate and assist policy-makers in identifying priorities for action and research.

3.

EXPERT PARTICIPATION

3.1. Selection of participants

Participating experts were required to have at least five years' experience in the field.

The experts were proposed by the Topic Centre Research on Work and Health (TCWH) members and the focal points of the Agency in order to ensure a broad coverage of qualified expertise across the EU. For their answers to be taken into consideration, the respondents had to have at least five years' experience in the field of biological risks.

The expertise was collected and used with full awareness of the principles and guidelines of the European Commission ([16]).

3.2. Responses

For the first round, 109 experts were approached by the TCWH and invited to participate in the survey. 32 experts returned completed questionnaires (response rate: 29 %).

109 experts were approached. The forecast is based on 36 questionnaires returned from 21 European countries.

Ninety-five experts were invited in the second phase, of which 42 returned completed questionnaires (response rate: 44 %).

The same number of experts (N=95) was invited to take part in the third and last survey round. Thirty-six questionnaires were returned (response rate: 38 %).

All the questionnaires received in the three rounds were returned from experts with at least five years' experience in the field of biological risks.

Over the three survey rounds, experts from 21 European countries participated in the formulation of the forecast on emerging OSH biological risks (Diagram 2).

It should be noted that one third of the answers were received from Germany, which is therefore over-represented in the survey. This may have biased the results towards the German position on biological emerging OSH risks, hence the forecast may not be representative of a European consensus.

([16]) European Commission, 'Collection and use of expertise by the Commission — principles and guidelines', Luxembourg, 2004, ISBN: 9289458216, http://europa.eu.int/comm/research/science-society/pdf/guidlines_ss_en.pdf.

Diagram 2. Number of respondents from different countries of origin completing the first, second and third rounds of the survey

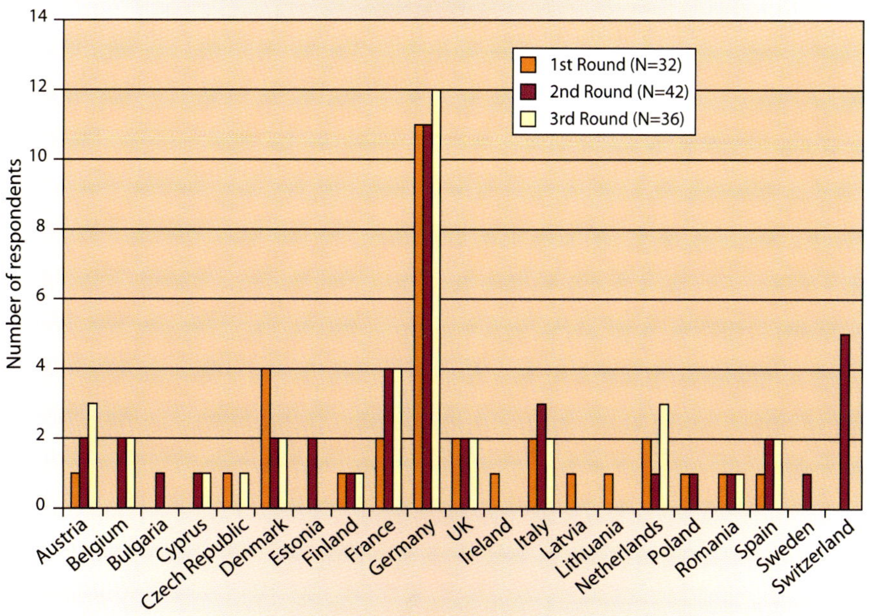

3.3. Characteristics of respondents

3.3.1. Functions of the respondents

Diagram 3. Number of respondents to the first, second and third survey rounds, by function

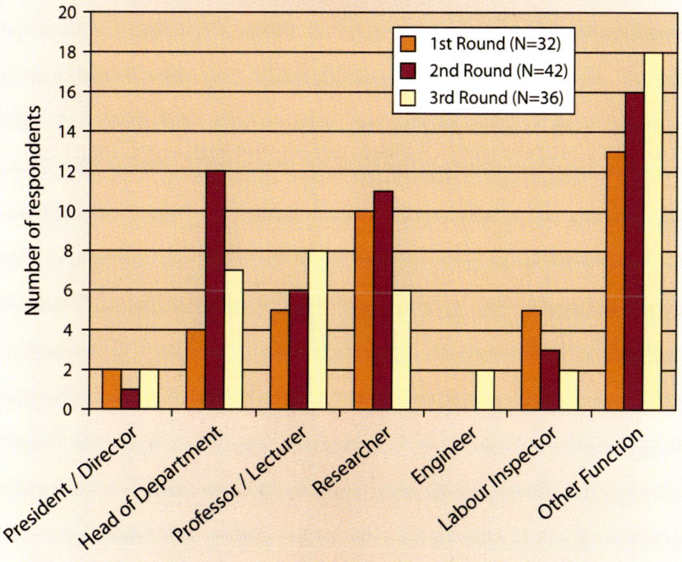

In the third survey round, which forms the forecast, the majority of respondents were 'professors/lecturers' (n=8), 'heads of department' (n=7) or 'researchers' (n=6). Eighteen experts (additionally) ticked 'other function'. These other functions are specified below: 'OSH medical counsellor' (n=3); 'safety advisor'; 'head of section'; 'specialist'; 'OSH technician'; 'expert'; 'biosafety manager'; 'biological safety officer'; 'university biological safety advisor'; 'head of department 'biological agents'; 'coordinator biotechnology regulations'; 'member of division and scientific expert' as well as 'biosafety manager'. Two of them did not specify their 'other function', but indicated that they were involved in inspection and consulting activities (Diagram 3).

3.3.2. Fields of activity of the respondents

Most of the respondents to the third survey round were involved in 'training/teaching' (n=16), in 'research' (n=15) or in 'consulting' (n=15).

Five experts ticked 'other main activity': Besides being involved in 'training' and 'consulting', three of them indicated they were 'biological risks — project managers', one specified 'national legislation' and another expert wrote that his/ her tasks 'involve most of the activities mentioned in the questionnaire'. All these activities were considered acceptable and all experts met the selection criteria defined (Diagram 4).

Diagram 4. Number of respondents to the first, second and third survey rounds, by field of activity

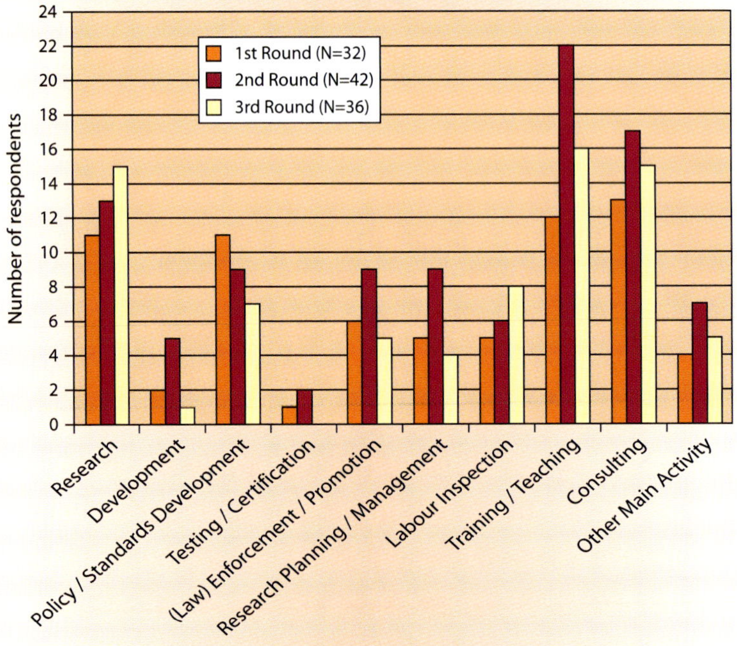

4.

MAIN EMERGING BIOLOGICAL RISKS IDENTIFIED

4.1. Survey results

The 10 main emerging risks highlighted in the forecast are presented in this chapter. The exact descriptions of the risks as rated by the experts are listed in Table 1 together with the number of respondents to each item, the mean value of the ratings and the standard deviation.

Diagram 5. The 10 most important emerging biological risks identified in the survey

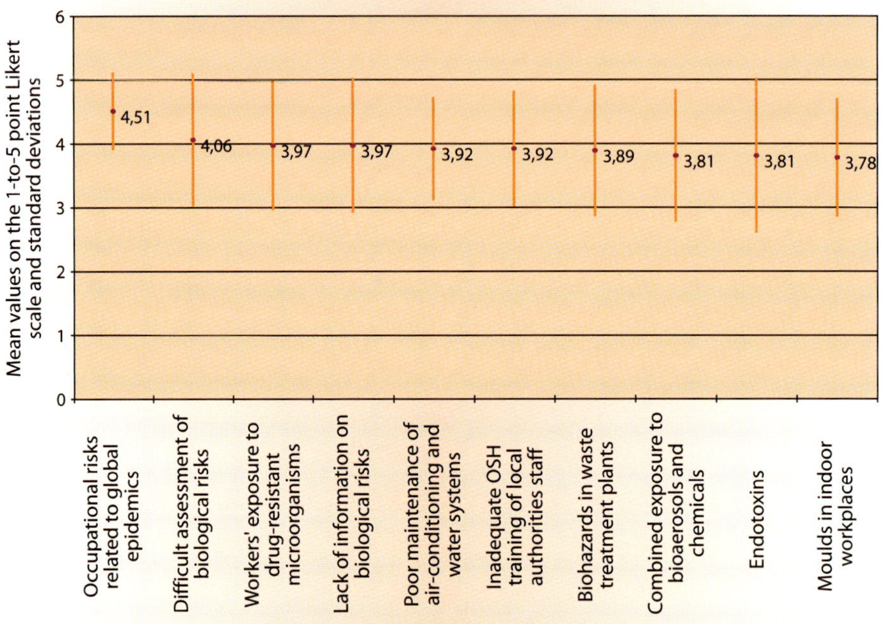

The forecast singles out two risks 'strongly agreed' as emerging (MV>4): the **'occupational risks related to global epidemics'** (MV=4.51); and the **'difficult assessment' of risks posed by biological agents in the workplace** (MV=4.06). However, it should be noted that the limit for distinguishing between the risks 'strongly agreed' and 'agreed' as emerging was set arbitrarily (see '2.1. Implementation of the expert survey') and that the mean value of the last item rated as 'strongly agreed' as emerging risk is very close to the mean values of the next following items.

The top risk also shows a high consensus among the experts (SD=0.612). Some examples of diseases mentioned by the experts are **severe acute respiratory syndrome (SARS), viral hemorrhagic fever, tuberculosis, acquired immune deficiency syndrome (AIDS), hepatitis C and hepatitis B**. But the list of emerging epidemics affecting the occupational environment in the context of the changing world of work is long. More information on these risks can be found in the experts' comments (see below) and in a literature review (see '4.2.1 Occupational risks related to global epidemics'). Occupational risks in the context of pandemics have also been identified as major OSH priorities in a review of various national, EU and international resources identifying future research needs in the field of OSH, carried out by the

Agency ([17]). The consistent results between the forecast and the review strengthen the importance of this issue.

It is interesting to note that the OSH risks linked to epidemics, together with the third of the 'top' emerging risks — the risk of workers' contamination with drug-resistant microorganisms (see experts' comments below and literature review '4.2.2 Workers' exposure to antimicrobial-resistant pathogens in the health care sector and livestock industry') — also relate to the public health sphere, hence the need to coordinate action between the OSH and public health areas to ensure efficient prevention. Although the latter field is clearly outside the remit of the Agency, it is essential to stress the importance of **multi-disciplinary cooperation and coordination**. Effective use and sharing of research and information is of the utmost importance.

Poor risk management is a major problem. Whilst according to Directive 2000/54/EC employers have the duty to determine and assess the risks posed by biological agents in the workplace, in practice, proper assessment of biological risks is still difficult (see '4.2.6 Difficult risk assessment of biological agents in the workplace for more information') and is strongly agreed to be an emerging risk in itself by the respondents to the survey (MV=4.06). The 'lack of information on biological risks' in the workplace (MV=3.97) contributes to render risk assessment difficult and belongs as well to the top 10 emerging risks. The consistency in the respondents' evaluation of several items linked with risk assessment (see more related items in '5.3 Biological risks resulting from poor risk management and prevention practices') may be considered to validate the forecast. Besides, it is noticeable that the problematic exposure assessment of biological agents has been identified as a priority in the review of OSH research priorities mentioned earlier.

Besides deficient risk management, inadequate or lacking preventive measures such as the 'inadequate provision of OSH training to workers' (MV=3.92) — especially in local authorities — or the 'poor maintenance' of equipment — water and air-conditioning systems being put forward here (MV=3.92) — are also considered to pose emerging risks to workers. Moreover, this illustrates the consequences of insufficient information on biohazards. Indeed, the experts comment that some ill-health symptoms observed in indoor workers are often wrongly assumed to be flu-like diseases; in fact, they are engendered by **biological agents developing in poorly maintained air-conditioning systems**. Recent findings on legionella should now facilitate establishing a correct diagnosis of such symptoms.

If the risks engendered by biological agents are difficult to assess, **'combined exposure to biological agents and chemicals'** (MV=3.81) is all the more challenging, and is actually closer to the reality of workplaces. While the range of potential health effects is wide, it is difficult to determine which of the constituents primarily accounts for which health effects.

The occupational risks linked to **'waste treatment'** are not only among the main emerging biological risks (MV=3.89) (see also literature review '4.2.5 Biological risks in the management of solid waste') but were strongly agreed to be emerging (MV=4.11) in a complementary expert survey on emerging chemical risks ([18]). The similar mean values in both surveys may be considered to validate each other.

([17]) European Agency for Safety and Health at Work, 'Priorities for occupational safety and health research in the EU-25', Luxembourg, 2005.

([18]) European Agency for Safety and Health at Work, 'Expert forecast on emerging chemical risks related to occupational safety and health'. The report will be published in 2007.

Last but not least, two of the main emerging risks put forward in the forecast are directly linked to specific biological agents, namely **'endotoxins'** (MV=3.81) (see literature review '4.2.3 Occupational exposure to endotoxins') and **'mould in indoor workplaces'** (MV=3.78) (see literature review '4.2.4 Moulds in indoor workplaces').

Table 1. The 10 most important emerging biological risks identified in the survey (N = number of experts answering the specific item; mean value (MV); standard deviation (SD))

- MV>4: risk strongly agreed as emerging
- 3.25<MV≤4: risk agreed as emerging
- 2.75≤MV≤3.25: status undecided
- 2≤MV<2.75: risk agreed as non-emerging

NB: None of the risks were strongly agreed as non-emerging (MV<2).

Top ten biological risks	N	Mean Value (MV)	Standard Deviation (SD)
Globalisation leading to epidemics of old and new pathogens (e.g. Severe Acute Respiratory Syndrome (SARS), avian flu, viral hemorrhagic fever, tuberculosis, Human Immunodeficiency Virus (HIV), Hepatitis C, Hepatitis B): • High density of animals in confined spaces in contact with humans leading to increasing zoonosis cases (diseases jumping the species barrier from animals to humans). • High population density and increase in business trips, tourism and immigration helping zoonoses and other infectious diseases to spread quickly worldwide. Groups particularly at risks of contamination: staff involved in producing, processing and transporting livestock; airport staff and air crews; staff involved in border controls and policing; staff in health care sector; public transport; and public services. The risk is often underestimated, which leads to a lack of preventive measures.	35	4,51	0,612
Poor or difficult assessment of biological risks.	36	4,06	1,040
General increased use of antibiotics for human health care and for animal breeding in the food industry leading to the apparition of drug resistant pathogens (e.g., methicillin resistant staphylococcus aureus (MRSA), tubercule bacillius (TBC)). Health effects observed: increase in staff infected with MRSA in western hospitals; increasing antibiotics resistance of livestock farmers and in the population in general.	35	3,97	1,014
Lack of information on biological risks in different workplaces (e.g. office workplaces, agriculture).	36	3,97	1,055
Poor maintenance of air-conditioning (whose use is increasing) and water systems (e.g. legionella, aspergillosis in hospitals). New knowledge about the presence of legionella will help the correct diagnosis of symptoms so far wrongly attributed to other diseases like flu.	36	3,92	0,806
Inadequate training, poor knowledge of OSH or even poor basic awareness of risks of local authorities staff (e.g. sewage, excavations, waste collection, etc.).	36	3,92	0,906
Biohazards in waste treatment plants (e.g. selective sorting, manufacture of compost) leading to allergies, infectious diseases (bacteria, viruses), toxinic diseases (endotoxins, mycotoxins) and cancers (oncogens). Especially in composting facilities where there is a wide variety of microorganisms present at the different stages of the composting process, the risks are not completely identified yet.	36	3,89	1,036

Top ten biological risks	N	Mean Value (MV)	Standard Deviation (SD)
Bioaerosols and chemicals, the combined effects of which have been very little studied but lead to allergies. More knowledge will help identify the real multi-factorial causes of symptoms for which mono-causal explanations have been given so far.	36	3,81	1,037
Endotoxins: High concentrations in various industrial settings (e.g. in workplaces exposed to organic materials (straw, wood, cotton dust), waste treatment, poultry houses, swine confinement buildings) leading to asthma, loss of lung function, etc.	36	3,81	1,215
Moulds in indoor workplaces due to new construction methods and materials, due to the aim of saving energy and due to the lack of maintenance: Exposure to fungal spores for office workers and especially workers involved in building restauration, leading to sensitisation and allergies.	36	3,78	0,929

Experts' comments

When available, the comments added by the respondents to the items are listed below to provide some context and support to the ratings.

Risks strongly agreed as emerging (MV > 4)

- Occupational risks related to global epidemics

There is a real risk of global epidemics of endemic diseases such as malaria, dengue fever, and of meningococcal disease and measles. The following groups of workers face increased risks of contracting such diseases in their jobs: health care staff, livestock handlers, airport staff and workers involved in border controls. Aircrews are also at risk as they are exposed to poorly filtered air recirculated into the aircraft cabins. Additionally, drivers in public transport are at risk of coming in contact with infected people and hence being contaminated. With regard to the livestock industry, close contact between human beings and animals in confined spaces is a long-established practice that has contributed to influenza pandemics through antigenic shifts for centuries. According to the respondents, the rise in global travelling is another factor responsible for the increasing risk of pandemics. In this view, the need for better information systems alerting travellers on these risks is emphasised. Generally, more research is needed to provide more detailed data on workers groups at risk in order both to help employers implementing preventive measures, and to give policy-makers evidence of the need for an increase in research funding.

- Poor or difficult assessment of biological risks

Proper assessment of biological risks is necessary for prevention. According to one respondent, the assessment of biological risks is usually better done in laboratories and in the health care sector. However, identification and measurement of biological agents in general are major issues and still need to be addressed.

Risks agreed as emerging (3.25 < MV ≤ 4)

- Workers' exposure to antibiotics-resistant bacteria in the human health care sector and in the food industry

The increased use of antibiotics for human health care and for animal breeding, as well as the inadequate use of antibiotics (for example, a dosage that is too low, a treatment that is not followed until completion, the choice of an antibiotic without a study on resistance etc.) leads to the emergence of drug-resistant bacteria, such as multiresistant tuberculosis. However, despite supporting evidence for this phenomenon, quantitative epidemiological information is rather weak. Therefore, more research has to be conducted in this field.

- Poor maintenance of air-conditioning and water systems

Workers involved in maintenance activities are at greater risk of exposure to legionella.

- Combined exposure to bioaerosols and chemicals

More knowledge on the negative health effects of the combined exposure to bioaerosols and chemicals is still needed.

- Endotoxins

High concentrations of endotoxins can be found in the agricultural sector (swine breeding and grain harvest). One issue that is often ignored is that the use of bactericides to eliminate bacteria from a contaminated area may actually result in the emergence of other endotoxin-producing organisms in the very same area, possibly resulting in endotoxins resistant to the bactericide used and in concentrations even higher than the initial level. However, besides their negative health effects, endotoxins may also have positive consequences. Indeed, exposure to higher concentrations of endotoxins may also reduce the incidence of allergic reactions such as atopic asthma and allergies through an effect on the balance of T-helper cells 1 and T-helper cells 2. More research is needed on occupational exposure to endotoxins and on the dose-effect relationship.

- Moulds in indoor workplaces

Due to financial considerations, constructors often do not leave building materials enough time to dry off, which may result in mould growth in the finished construction. Indoor mould growth is also found in older buildings. According to one respondent, moulds are only moderate allergens, and only high airborne concentrations of mould spores may cause allergies. A further expert adds that mould-related allergies in workers seem to be an issue for construction workers rather than office workers. However, another expert mentions that an increase in the number of mould-related occupational diseases is seen in some countries' statistics and adds that in Finland in 2002 there were 264 cases of occupational diseases caused by moulds, mostly allergies (155 cases). The most common branch for these occupational diseases was health care and social sector with 71 cases, followed by public administration (49 cases), agriculture (43 cases) and education (42 cases). Construction branch had only seven cases of occupational diseases caused by moulds. This divergence of opinion may be due to geographical differences, to the different level of awareness for mould-related health problems in the different countries, and to the differences in national recognition systems for occupational diseases. Last but not least, exposure to mouldy working environments due to water-damaged constructions may be the cause for health symptoms sometimes mis-diagnosed as flu-like diseases.

4.2. Literature reviews

This section contains six literature reviews that explore in more depth some of the main emerging risks singled out in the forecast in terms of context, workers at risk, health and safety outcomes and prevention:

- occupational risks related to global epidemics
- workers' exposure to drug-resistant microorganisms
- occupational exposure to endotoxins
- moulds in indoor workplaces
- biohazards in waste treatment activities, and
- the difficult assessment of biological risks.

The papers selected for these reviews all originate from scientific peer-reviewed journals, from reputable research or OSH organisations, or from conference proceedings reviewed by scientific committees. Out of the 366 references used, only 42 (11%) were published before 1996, i.e. more than 10 years before this report.

4.2.1. Occupational risks related to global epidemics

In the 21st century, we are still faced with the continuous emergence of new or newly recognised pathogens (such as the severe acute respiratory syndrome (SARS), avian influenza, Ebola and Marburg viruses) and the re-emergence of well-characterised outbreak-prone diseases (such as cholera, dengue, measles, meningitis and yellow fever). Whilst until the end of 2005, only three diseases — cholera, plague and yellow fever — had to be reported to the World Health Organization (WHO), the new International Health Regulation now encompasses all 'public health emergencies of international concern, including emerging diseases', which must be reported as soon as possible to the WHO[1][2].

In a joint consultation of the WHO, the Food and Agriculture Organization of the United Nations (FAO) and the World Organisation for Animal Health (OIE)[3], the following key zoonotic diseases have been identified for Europe:

More than three quarters of human diseases are zoonoses.

- zoonotic agents for which emergence will have a major impact on public health: avian influenza virus, as well as drug-resistant and more virulent strains of food-borne bacteria;
- zoonoses and zoonotic agents with current and potentially increasing impact: transmissible spongiform encephalopathies (TSEs), Hanta virus, rabies (Eastern Europe), orthopox virus, tick-borne encephalitis, hepatitis E (porcine), Lyme disease, Rickettsia spp., tuberculosis (bovine/avian), tularaemia, Brucella melitensis, marine brucellosis, Echinococcus multilocularis, Echinococcus granulosus, Leishmania spp., Taenia solium, trichinellosis, Baylisascaris ascaris (larval migrans), toxoplasmosis and cryptosporidiosis/giardiasis;
- zoonoses and zoonotic agents imported from outside Europe: Rift Valley fever, dengue virus, West Nile virus, alpha viruses, TSEs, pandemic influenza, SARS coronavirus, monkeypox, paratuberculosis, Borna virus, pathogens transmitted via blood and blood products, pathogens from marine environments (Vibrio spp., influenza A/B, Calici virus, Brucella spp., nematodes) and Burkholderia pseudomallei (potentially).

More than three quarters of the human diseases that are new, emerging or re-emerging at the beginning of the 21st century are zoonoses, i.e. caused by pathogens originating from animals or from products of animal origin. Many of these emerging diseases spill over from natural wildlife reservoirs into humans, either directly or via domestic or peridomestic animals. Emerging zoonotic diseases are increasingly recognised as a global issue with potentially serious impact on human health effects. Their current upward trends are likely to continue[3].

A new contagious pathogen can spread very rapidly around the globe.

When a contagious virus emerges, its global spread is considered inevitable and is likely to be rapid. The pandemics of the previous century encircled the globe in six to nine months, even at a time when most international travel was by ship. Today, it is estimated that a new contagious virus could reach all continents in less than three months. For example, SARS spread quickly to 30 countries, even though it was not caused by a highly infectious pathogen. A combination of factors may contribute to the emergence and spread of (zoonotic) disease agents, including:

- the increased speed and volume of international transport of humans, animals, and products;
- agricultural expansion and intensification, as well as new practices in animal husbandry and in food production responding to the increasing demand for animal protein;
- demography and the increasing number of immuno-compromised people as a consequence of an ageing population;
- ecological factors such as climate change, as well as urbanisation, land use and deforestation, whereby the contact between humans, domestic animals, and wildlife is increased, creating more opportunities for animal-to-human transmission of diseases; and
- factors associated with the disease-causing agent, such as the development of increased virulence or drug resistance, adaptation to new vectors and hosts, as well as mutation and recombination in humans and other animals after exposure to multiple pathogens (for example, food-borne or influenza viruses).

Measures such as border closures and travel restrictions might delay but cannot stop the introduction of a microorganism into a country; for example, it is not possible to stop the migration of wild birds, which may spread a disease-causing agents over national borders[4][5][6].

To date, research into biological hazards in the occupational environment remains limited apart from the risks of infectious disease affecting both research and health care professionals. Moreover, the extensive use of antibiotics and vaccines has resulted in less surveillance of the risks of acquiring infections and transmissible diseases. However, the emergence of the Human Immuno-deficiency Virus (HIV) in the 1980s alarmed infectiology experts and rapidly affected the world of work. The new virus originated in Africa and presented new occupational risks for all health care workers, requiring changes in medical practices and new safety regulations for handling blood and other biological fluids. Such measures, however, do not protect workers from airborne transmitted infection, such as SARS or avian flu. Most of the respiratory masks commonly available on the market do not provide efficient protection, while medical masks do not protect sufficiently against bioaerosols. A great deal remains to be done in this field.

Global epidemics and the epidemiological chain

The internationalisation of traffic and trade may have an impact on two of the five links of the epidemiological chain(19), which is also called the transmission chain: the reservoir link (the source) and the transmission-by-vector link leading to the introduction of a new epidemiological cycle on a previously epidemic-free geographic area. The reservoir and the vector may be an animal — for example, wild birds are a suspected reservoir of avian flu, and the mosquito *Aedes albopictus* is a vector of malaria, dengue, yellow fever, etc. — or a human being[3].

Action of globalisation on the reservoir link

Livestock as a reservoir

Today, avian flu is a major concern with the threat of a pandemic flu. The highly pathogenic avian influenza (HPAI) viruses with most significant impact on humans arise from influenza A strains, such as A/H5 or A/H7. The natural reservoir of these 'A' strains is found in aquatic wild bird populations. However, outbreaks of HPAI rarely start directly in wild birds, but are generally the result of some low pathogenic A strains being transmitted from wild birds to domestic poultry, among which they then circulate and eventually mutate into HPAI viruses.

Extensive poultry farming

Since 2000, there have been more and larger outbreaks of HPAI in poultry. This phenomenon is not yet understood, and is the subject of speculation. In particular, very large outbreaks have occurred in densely populated commercial bird populations, such as the A/H7N1 outbreak in Italy in 1999 and the A/H7N7 outbreak in The Netherlands in 2003[7]. In a joint consultation on emerging zoonotic diseases

(19) The epidemiological chain (medical term) or transmission chain (technical term preferred by workers) is the basis of an occupational biological risk assessment. It is composed of five steps; the reservoir (human, animal or non-living), the ways out, the transmission routes (direct, semi-direct via hands, indirect by vector), the entrances (skin, mucous membrane, airways, etc.) and finally the host — here the worker at the work place.

organised by the WHO, 89 persons working with infected poultry were reported to have been infected with A/H7N7 during the Dutch outbreak[3]. According to a study published by the Dutch National Institute for Public Health and the Environment (RIVM)[8], of the 500 tested persons who had handled infected poultry in the Netherlands, about 50% showed an antibody response. Antibodies were also found in 59% of infected poultry workers' family members.

Globalisation of trade and traffic has contributed to the spreading of avian influenza.

To date, outbreak centres of avian influenza A/H5N1 have multiplied and spread to about 50 countries on three continents (Asia, Africa and Europe). While wild migratory birds generally tend to be seen as the factor responsible for the spreading of the epizootic, globalisation of trade and traffic certainly contribute to this phenomenon[9][10][11][12]. Indeed, in 2005, two raptors illegally introduced in Belgium were infected with the avian influenza virus A/H5N1[13]. In Northern Asia and in Siberia, the diffusion of the A/H5N1 virus has been found to follow the railways[14]. In Nigeria, the introduction of the A/H5N1 virus into an intensive poultry-breeding industry is thought to be the result of trading activities with affected countries[9][10][11][12]. Further widespread infections in domestic poultry have been reported in parts of north, west and central Africa. The most important drivers are likely to be trade and movement of poultry, rather than wild birds. In the case of the well-described outbreaks in Egypt, both commercial and backyard flocks were involved[15]. At the other extremity of the epidemiological chain, workers are often the very first ones to be exposed to such new infectious agents. More than anyone else, poultry workers face intensive, daily occupational exposure to the risk of contamination with the avian influenza virus(17)(see below 'Exposure routes and workers at risk').

Exotic animals and imported reservoirs

The fashion for exotic domestic pets exposes international trade workers to zoonosis.

In the northern hemisphere, the new fashion for exotic domestic pets has encouraged the import of rodents, bats and reptiles, introducing some opportunities for exotic or classical zoonosis to develop in workers involved in this trade. Two and three years ago, pups from Morocco and Egyptian bats brought into Europe introduced rabies[16]. In 2004, the cross-breeding of African rodents and prairie dogs in an American animal housing facility resulted in an epidemic of monkey-pox with more than 80 people infected. Hundreds of prairie dogs had been in transit in this animal housing facility and all owners, importers and sellers in the United States had to be traced[17][18][19]. In the southern hemisphere, deforestation has increased contact between human populations and small wild animals such as rodents and bats. Workers from the northern hemisphere trading directly with these people are exposed to the same local zoonosis (Ebola fever, for example)[4][20].

Human reservoirs as a consequence of long-distance travelling

Increasing international travel and air travel multiply the opportunities for pandemics.

Besides international trade, the increase in international travel and more particularly in air travel as a consequence of more frequent business-related trips and of increased tourism has multiplied the opportunities for new epidemics. In this case, human beings act as a reservoir in a geographic area previously free from the pathogen. Beginning in China at the end of 2002, the international spread of SARS resulted in 8,098 SARS cases in 26 countries, with 774 deaths by July 2003[21]. This new Corona virus also affected the occupational environment, since SARS cases were found in workers in contact with contaminated individuals, such as airport and health care staff. The WHO alerted the global community to SARS epidemics among health care workers in Hanoi, Vietnam, and Hong Kong. SARS was also introduced to Toronto in 2003 via an infected individual coming back from China, which led to a total of 128 further SARS cases, most of them in staff of the hospital where the initial case was treated. This episode resulted in the death of three health care workers, two nurses and a doctor. In some cases, contaminated staff brought the diseases into their homes, which led to secondary contamination of family members. Furthermore, the WHO alerted the global community to a severe respiratory syndrome that spread among health care workers in Hanoi, Vietnam, and Hong Kong[22][23]. Since its emergence, the very origin of the SARS Coronavirus (SARS-CoV) has proved elusive. New research findings suggest that bats may be a natural reservoir: it has been found that species of bats are a natural host of SARS-like coronaviruses (SL-CoVs), which are closely related to those responsible for the SARS outbreak. As the human isolates of SARS-CoV nestle phylogenetically within the spectrum of SL-CoVs, it means that the virus responsible for the SARS outbreak was a member of this SL-CoVs group[24]. If bats are the actual reservoir of the SARS-CoV, this raises important questions about how to monitor and control emergent disease outbreaks. More particularly, further study of bats is warranted, particularly since bats are sold in live-animal markets and consumed by humans in China.

Several hospitals in the European Union and in Switzerland have been alerted to, or have dealt with, cases of imported haemorrhagic viral fever. It is not possible to draw up a complete list, but it seems that between 1980 and 2000, there were five confirmed cases of Lassa fever and one of Ebola haemorrhagic fever, as well as seven suspected cases of Lassa fever and five of Ebola haemorrhagic fever. In 2004, a patient with Crimea-Congo haemorrhagic fever was sent back to France from Senegal with an incorrect diagnosis. No special precautions were taken and it was later calculated that the patient had been in contact with 181 people, including 97 staff at the French hospital and two privately employed ambulance drivers. Also in 2004, the United States reported one case of Lassa fever in a businessman returning from West Africa. A total of 139 health care workers and 16 private laboratory workers had been in contact with him[25][26][27][28][29][30][31][32][33]. Fortunately, to date, no secondary clinical cases of Crimea-Congo haemorrhagic fever or Lassa fever have been reported, meaning that the epidemiological chain could be broken after introduction of the reservoir.

Action of globalisation on the transmission link

Over centuries, urbanisation and the development of transport have modified the establishment of vector-transmitted diseases by encouraging both the dispersal and establishment of vectors[34][35][36]. Today, the increasing speed and number of exchanges multiply the risks of importing vectors of tropical diseases.

Workers handling international trade containers are at risk of mosquito-borne Dengue fever, a major international public health concern.

Imported goods, particularly in water, may introduce disease vectors such as mosquitoes and other arthropods. Dengue, for instance, which is transmitted by the main vector Aedes aegytpi and Aedes albopictus mosquitoes, is the most common mosquito-borne viral disease affecting humans and has become a major international public health concern in recent years. Over the past 25 years, the geographical spread of both the mosquito vectors and the viruses has led to the global resurgence of epidemic dengue fever and to the emergence of dengue hemorrhagic fever (DHF). While only nine countries were affected by DHF in 1970, the number has increased more than fourfold and continues to rise. It is estimated that dengue affects about 50 million people and kills about 24,000 people every year[37][38]. Geographical expansion of the mosquito has been aided by increased air travel and by international commercial trade, particularly where water is imported either unintentionally — for instance in used tyres which easily accumulate rainwater — or intentionally in the trade of Dracaena sanderiana (so-called 'lucky bamboo') imported in water[34][37][39][40]. In Europe, Aedes albopictus appeared in Albania in 1979, in Italy

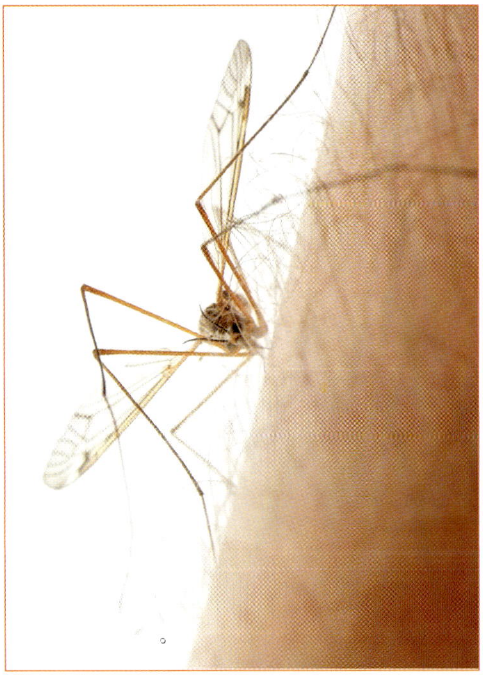

Mosquito

in 1990, in Belgium in a large tyre-import plant in 2000, and in Serbia and Montenegro in 2001[38][40]. In France, it was first identified in 1999 in a large plant recycling tyres imported from the United States and from Japan. Since then, Aedes albopictus has been found in several tyre-import plants throughout France[40]. The international trade in used tyres has been found to be particularly implicated in the dispersal of certain types of mosquitoes with an ecological plasticity that enables them to adapt to new latitudes[9][10][11][12][13][23]. In North America, Aedes albopictus was found in 2001 on a cargo ship at Los Angeles harbour with lucky bamboo in water coming from China[38]. It was then found in 14 resellers in California. There is a real risk of infection in workers exposed to the mosquito-vector if the imported mosquito is already infected with the virus at the time of its importation. In the field, it is not possible to distinguish already infected from non-infected Aedes albopictus. Prevention measures should be systematically adopted whenever there is a risk of introduction of exotic mosquitoes through importation. In France, for example, the risk of its spreading is currently being monitored very closely. Last but not least, Aedes albopictus may also be a vector for other diseases such as malaria, yellow fever, West-Nile fever, Japanese encephalitis and St Louis encephalitis[36][39][40][41].

In 1967, outbreaks of Marburg haemorrhagic fever — a severe and often fatal disease caused by a virus from the same family as the one that causes Ebola haemorrhagic fever — in Germany and Yugoslavia have been linked to laboratory work using African green monkeys (*Cercopithecus aethiops*) imported from Uganda. The outbreaks involved 25 primary infections in laboratory workers, with seven deaths. Among the six secondary cases that followed, two occurred in doctors and one in a nurse. A new epidemic, which began in 2004 in Angola, is still ongoing. Despite years of intensive investigation no animal reservoir or other environmental source of the Marburg virus — or of the Ebola one — has been identified. Although monkeys are susceptible to infection, they are not considered to be a viable reservoir but instead act as a vector[42][43].

Exposure routes and workers at risk

In the context of global risk, it is not easy to identify the at-risk occupations since sources of exposure are varied and involve people, plants, goods and animals.

In the case of human-to-human transmission, health care workers[22][23][44] for example are at the front line of the contamination risk. According to the 'World health report 2002 — Reducing risks, promoting healthy life'([20]) by WHO, among the 35 million health care workers worldwide, about three million are exposed to blood

([20]) http://www.who.int/whr/2002/chapter4/en/index8.html

borne pathogens each year via percutaneous injuries — 2 million to hepatitis B (HBV), 0.9 million to hepatitis C (HCV) and 170,000 to HIV. These injuries may result in 70,000 HBV, 15,000 HCV, and 500 HIV infections. More particularly with regards to HIV, the risk of contamination of health care workers is almost exclusively related to needle stick injury. The probability of HIV infection following a needle stick injury is 0.32% when the source is HIV-positive [23].

Handling blood samples in the health care sector

Health care workers, sex workers and transport workers are exposed to the risk of HIV contamination.

However, not only health workers are exposed to HIV. By the end of 2005, an estimated 24.5 million workers were living with HIV out of the 38.6 million persons infected worldwide — so the ILO's report 'HIV/AIDS and work: global estimates, impact on children and youth, and response — 2006'(21). The 'ILO code of practice on HIV/AIDS and the world of work'(22) established in 2001 recognises HIV/AIDS as a workplace issue not only because it affects the workforce but also because the workplace can play a crucial role in limiting HIV transmission. A lack of opportunities for decent work can compel people to work under precarious and un-regulated conditions where they may be at increased risk from HIV(21). The World Bank, in the report 'Combating HIV/AIDS in Europe and central Asia'(23), warns for example against the explosion of commercial sex work, which represents a major threat in eastern Europe and central Asia where HIV/AIDS epidemics is one of the world's fastest-growing. Generally, young women face greater risks than their male counterparts of being sexually abused and acquiring HIV at their workplace, particularly through prostitution and other sexual exploitation(21).

Other groups of workers, such as transport workers and more generally mobile workers, are particularly vulnerable to HIV/AIDS — and other sexually transmitted infections — due to the particularity of their working conditions. Transport workers encounter for instance few recreational opportunities while on the road and may be tempted to compensate this with alcohol and prostitution. Additionally, clean and

(21) http://www.ilo.ru/news/200612/Global_Estimate.pdf

(22) http://www.ilo.org/public/english/protection/trav/aids/publ/code.htm

(23) http://www.aidsmedia.org/files/985_file_World_Bank_English.pdf

secure sleeping accommodations at truck stops can be expensive and some drivers report that spending the night with a commercial sex worker is sometimes cheaper. In eastern Europe, a survey of truck drivers revealed that more than 80 per cent of them had spent more than four months away from home the previous year, with 36 per cent indicating they had had casual sex. In some locations frequented by transport workers condoms are not always available or sometimes very expensive ([24]).

Workers in contact with live or dead infected animals, aerosols, dust or surfaces contaminated by their secretions are at risk of zoonoses.

Regarding animal-to-human transmission of zoonoses, workers in contact with live or dead infected animals, or with aerosols, dust or surfaces contaminated by their secretions have a higher risk compared to the general population. It includes: workers in legal or illegal animal trade; workers in farms, animal husbandries, breeding facilities and slaughtering facilities, including those involved in the disposal of carcases and the cleaning and disinfection of contaminated areas; workers in veterinary services; as well as in research[7][13][16][17][45][46][47][48][49][50]. Other workers also exposed to animals, such as custom officers, zoo staff, pet shop workers and gamekeepers may also be at risk[13][17][51].

Air travel plays a key role in the spread of communicable diseases over international boundaries. Air travellers in general — including those travelling for their job and more particularly air crews — are at risk of being contaminated with diseases transmissible via aerosols, such as SARS[22] or tuberculosis[52].

Those working to control epidemic outbreaks are another obvious risk group[7]; for example, workers involved in the culling and disposal of herds infected with avian flu and bovine spongiform encephalopathy (BSE)[45][46][49][53].

Air crews and travellers, those working to control epidemics, media professionals and workers in war zones are also at risk of communicable diseases.

Media professionals in radio and television are also at risk[54]. In 1999, for example, one fatal case of yellow fever was found in Germany in a non-vaccinated cameraman working on location in equatorial Africa[55].

Whatever the goods, all activities related to imports may be implicated in the introduction of vectors such as mosquitoes. Here, there is a risk of contamination with malaria or dengue fever, for example. All professions involved in these activities are at risk, especially those unloading and opening containers. Indeed, cases of malaria have been found in airport staff who had not been abroad for many months[56].

International trade — handling containers

War, peacekeeping operations or the distribution of humanitarian aid may also spread diseases or their vectors[57]. WHO reports that several 'old' and well-known zoonotic diseases appear to be re-emerging in the WHO European region, as a result of war among other reasons. For instance, the WHO Regional Office has been particularly involved in dealing with leishmaniasis in HIV-immunodepressed people in Italy and Spain as a result of imported cases from Afghanistan[3].

In the case of SARS, transmission occurs mainly through respiratory droplets and direct contact[23][58][59][60], but also through indirect vehicle-borne transmission;

([24]) http://www.ilo.org/public/english/dialogue/sector/techmeet/tmrts06/report.pdf

for example, via contaminated materials. Indeed, the strain of the Corona virus causing SARS can live on surfaces for at least two days and, in moist conditions, for up to four days[22]. Cases of transmission have been found mostly in individuals who had had close contact with a SARS-infected individual, including health care staff. In a study of nosocomial(25) transmission in Toronto, the highest infection rate of 60% was observed among health care staff in the coronary care unit where intense, close-contact care is given to patients[23].

Avian influenza

With regard to avian flu, research on genomic approaches, animal models and recombination approaches currently aims to determine which characteristics allow viruses to infect humans[7]. Indeed, despite recent progress, knowledge of the epidemiology, natural history, and management of influenza A/H5N1 disease in humans is incomplete. There is an urgent need for more coordination in clinical and epidemiological research among institutions in countries with cases of influenza A/H5N1 and internationally. In addition, seroprevalence studies are also urgently needed to determine the frequencies of human infection[48]. The infectious doses to humans are not known[45], but transmission probability depends on both virus and host factors[7]. With regard to the severity of the disease in the 140 or so cases of H5N1 human infections reported in Asia during 2004 and 2005, mortality was around 50%[7][61]. After a two to 14-day incubation period, fever and cough symptoms appear and, in almost all cases, end in pneumonia. In contrast to other human flu viruses, the A/H5N1 virus may spread out of the respiratory tract and harm other organs, such as the liver, kidneys or bone marrow[61].

Knowledge of the epidemiology of influenza A/H5N1 in humans is still incomplete but transmissibility from animals to humans seems very low.

Transmissibility of A/H5N1 from animals to humans seems very low, even for those directly exposed. Indeed, in Asia very few infections in humans have taken place in spite of the massive exposure. However, anti-H5 antibodies have been found in workers who have been exposed intensively to sick poultry, which suggests an increased risk for avian influenza infection from occupational exposure[62]. Prolonged exposure or close, direct contact with live or dead infected birds or poultry, to their infected tissues, excretions, or secretions — especially saliva, faeces and respiratory secretions — have been identified as the main risk factors. Transmission from animals to humans may occur[7][45][47][49]:

- via the respiratory tract, following the inhalation of fine dust or fine water droplets;
- by projection of fine contaminated dust onto the ocular mucous membranes;
- by hand-to-mucous membrane transfer (especially the ocular and nasal mucous membranes); for example, following the contamination of hands with surfaces contaminated with animal secretions.

Occupations at risk from avian flu include workers at different stages along the food chain, from farming to processing and food preparation.

At-risk occupations therefore include workers at different stages along the food chain, from farming to processing and food preparation. This includes workers who have direct contact with live or dead contaminated poultry, such as workers who breed, handle and slaughter poultry, clean slaughtered poultry, dispose, transport and destroy poultry carcasses, or handle uncooked poultry carcasses and handle uncooked meat in the food industry[46][47][48][49]. The risk of contamination is higher in containment areas (for example, in husbandries, birdhouses or at animal markets), or when people live very close to animals, as is sometimes the case in rural areas of South-East Asia[46]. However, since most of Europe's poultry flocks are

(25) Nosocomial infection: infection acquired in an hospital that was not present or incubating prior to the patient being admitted to the hospital, but occurred within 72 hours after admittance to the hospital. Taken from: http://www.medterms.com/script/main/art.asp?articlekey=4590

segregated from humans — unlike in Asia — the risk is low in this region[7]. The European Centre for Disease prevention and Control (ECDC) warns that the greater risk for human infection would be exposure to poultry raised or kept outdoors and which have contact with migratory birds. Indeed, poultry are highly susceptible to A/H5N1 viruses, therefore making the probability of exposure, infection, amplification and human infections greater through contact with outdoor-reared poultry than with indoor-reared commercial or industrial poultry[7].

Workers who are not in direct contact with poultry but come into contact with contaminated surfaces and dust are also at risk; for example, those cleaning or disinfecting breeding and hatching facilities or slaughterhouses, workers in vehicles used to transport poultry, or staff in poultry processing plants[45][46][47][49]. Since eggs seem to be a surface most likely to retain traces of faeces, workers handling, cleaning and processing eggs must be added to the list of jobs at risk of occupational exposure to the H5N1 virus[47]. Given the survival of influenza A/H5N1 in the environment, farmers may also be at risk through the widespread use of fertilisers derived from poultry manure[47][48]. At the other end of the food chain, hotel, restaurant and catering workers who handle uncooked chicken products are also at risk[47]. Last, but not least, at-risk occupations also include veterinarians or staff in diagnostic laboratories and veterinary research performing autopsy, handling smear samples and carrying out research on viral strains[45][46][49].

European researchers study the avian influenza viruses in order to develop new vaccines — Istituto Zooprofilattico Sperimentale delle Venezie, Italy

Human-to-human transmission is seen as even less likely than animal-human transmission.

Moving on to human-to-human transmission of the H5N1 virus, one study found H5N1-specific antibodies in the blood samples of eight out of 217 health care workers who had been exposed to contaminated patients. It concludes from epidemiological evidence that H5N1 viruses can be transmitted from patients to health care workers[44]. However, human-to-human transmission is seen as even less likely than animal-human transmission and, although three cases of human-to-human transmission have been reported during previous epizootic episodes, there has so far been no efficient person-to-person transmission. Those who have become infected were generally blood relatives providing care at home and, apart from one case, there have been no onward transmissions to those providing care in a health care setting who have taken normal precautions or among those controlling the disease[7][45][48][49].

This contrasts with the experience of other HPAIs where people working to control the disease have been more at risk. Hence, most evidence indicates a difficult adaptation process for A/H5N1 viruses among humans. However, a real concern is that a 'normal' seasonal flu virus infects an H5N1-infected human and that the two viruses recombine into a new efficient H5N1 strain. Seasonal influenza vaccination is therefore strongly recommended for workers involved in control measures when seasonal influenza is circulating. Currently, due to the low number of H5N1-infected humans, this would be statistically unlikely. However, the risk increases as the epizootic continues, especially considering the fact that other mammals may act as the 'vessel' for dual infection and recombination[7]. Besides, because large numbers of people in contact with poultry are likely to be at risk of H5N1 infection in African countries, if the virus recombines into a more efficient strain, then there is a possibility that a pandemic could arise from Africa[15].

Possible preventive measures

The very first levels at which intervention should be applied are the reservoir and transmission-vector levels. In this regard, the International Health Regulation regulates the loading of containers and transport vehicles to try and prevent the introduction of (known or new) pathogens or their vectors into a hitherto non-endemic area. These regulations, primarily concerned with public health, are also effective occupational prevention measures for workers involved in the import of containers and vehicles[1][2]. Furthermore, as many of the emerging infectious diseases are zoonoses[5], cooperation between public health and veterinary services must be reinforced. Trade of animals or animal products must undergo strict control.

At the workplace level, there is an urgent need to protect workers, in all sorts of jobs, from a new risk related to global epidemics. This is all the more so, given the difficulties in finding adequate treatment for some diseases[63]. Organisational, collective measures — including workers' training to identify at-risk situations and to apply the adequate control measures to both protect themselves and to stop the risks of further spreading — are necessary and should be complemented with the necessary personal protective measures.

In the case of the Canadian SARS episode, for instance, the precautions implemented were successful in halting the transmission in the hospital. Following the identification of staff and patient cases, the hospital was closed to admissions and discharged patients were placed into quarantine at home for 10 days. Along with an increased emphasis on hand-washing, additional precautions, including the use of gowns, gloves, N95([26]) or equivalent respiratory protection, and eye protection, were implemented for all patient care. Single or negative-pressure rooms were used for all febrile patients. Dedicated equipment was used for all patients, and patients were restricted to their rooms except for medically necessary tests. Staff wore N95 masks at all times in the hospital and were only allowed to leave their homes to go to work. Volunteers and medical students were excluded from the hospital, and patients' visits were restricted[23].

([26]) An N95 respirator is one of nine types of disposable particulate respirators. It filters at least 95 % of airborne particles but it is not resistant to oil. Taken from: http://www.cdc.gov/niosh/npptl/topics/respirators/disp_part/default.html

Pandemic preparedness plans are a prerequisite for responding adequately to threats at Community level.

Specific guidance for workers potentially at risk of avian flu is already available.

On the 28th November 2005, the Commission adopted an EU generic preparedness planning report[64] aiming to address public health threats and emergencies which are affecting or are likely to affect more than one Member State, whether anticipated (such as pandemic influenza) or unexpected (for example, a SARS-type epidemic)[65]. The existence of regularly updated national preparedness plans in the Member States is a prerequisite for responding adequately to threats at Community level. More specifically, with a view to meet the influenza pandemic, national influenza pandemic preparedness plans are being drafted or are already in place in all Member States[66][67]. Specific guidance for workers potentially at risk of avian flu is already available from the ECDC[7][10], from the WHO[68] and from some European national authorities(27) for workers in the livestock industry, poultry keepers, poultry slaughterers, workers involved in the disposal of poultry carcasses, workers in the food-manufacturing industry, health care and veterinary staff, for workers involved in outbreaks control, journalists in endemic areas and travellers including those travelling for professional reasons, etc.[7][45][46][54][69][70][71]. For example, in the health care sector, 'extended standard hygiene measures' are recommended in the case of exposure to a suspected case of a contaminated patient. In addition to the general hand-protection measures (such as wearing protective gloves and disinfecting hands), protective clothes, respiratory protection (such as FFP3-mask(28)) and goggles should be worn, especially when there is a possible formation of aerosols, such as when performing a bronchoscopy. Suspected patients should be isolated in single rooms. For staff who are directly involved in establishing a diagnosis or in therapeutic measures on confirmed or suspected cases, as well as for staff in direct contact with confirmed or suspected cases showing respiratory symptoms, it is recommended to take oral prophylaxis with a neuraminidase-blocker before or after exposure, depending on the situation. In order to avoid double infection, only health care staff vaccinated against seasonal flu should be permitted to attend to H5N1-patients[46].

Although the use of personal protective equipment (PPE) must be seen as the last resort, there is an urgent need for the development of efficient, adequate PPE against infectious risks. Indeed, the testing of most of the surgical masks commonly available on the market has shown that these do not protect adequately from airborne pathogens and infectious agents[72]. Furthermore, the clothing of some workers at risk, such as customs officers, does not protect them from the biological risks related to their occupation (for example, exposure to infected animals or people).

Broader organisational measures 'spilling over' into non-OSH domains should even be envisaged, such as revising the architecture of hospitals. Large hospital units may need to be restructured into smaller, geographically independent units with separate equipment. Unit isolation by controlled ventilation systems should be reconsidered, as it has proven ineffective in many situations[23][73]. Some experts also recommend major changes in the way the livestock industry is managed[74].

Last but not least, because pandemics pose a global treat, they require a global response. The SARS epidemic has demonstrated how sharing information and

(27) Collection of good practice information from various countries, European and international organisations available on the website of the European Agency for Safety and Heath at Work: http://osha.eu.int/good_practice/risks/dangerous_substances/index_topic?topicpath=/good_practice/risks/ dangerous_substances/bio_agents_zoonoses/

(28) The classification of available filtering half masks is carried out according to European Norms (EN 149) (Filtering Face Piece = FFP available in three classes P1, P2 and P3 providing differing protection factors (levels). (efficiency: low, med, high)

transparency are vital if infectious risks are to be fought. It has exemplified the role the international response can play in managing a global problem. Information sharing and management were coordinated by the WHO, and timely information was posted on its web site. Different international teams from various disciplines shared information and worked together to provide tools and results to enhance diagnosis and epidemiological analysis of the new pathogen[3]. More generally, particular attention should be given to new ways of cooperation, as well as the effective use and sharing of information[3][63]. The WHO has set up an international network using new diagnostic and communication technologies[75]. Europe has similar networks; for example, for legionellosis[76] and haemorrhagic fevers[77]. However, these networks currently barely take into account any occupational risk dimension, hence the urgent need to integrate occupational safety and health organisations among their members. Better collaboration is required both at horizontal level — between public health authorities, OSH authorities, veterinary public health services, food safety authorities, environmental protection authorities, wildlife agencies and social partners etc. — and at a vertical level — i.e. local, regional, national, and global — involving several disciplines, such as medical, veterinary, population biology, epidemiology, diagnostics, OSH, information technology, economics and social science etc.

The cooperation between various authorities is necessary for the effective prevention and control of epidemics.

The effective forecasting, surveillance, prevention and control of new emerging diseases is difficult given the complexity of the interactions between agents, animal host species and the environment, and considering the episodic and sometimes erratic nature of outbreaks. However, although history shows that the sequence of events leading to the emergence of a new disease is different each time, several factors are known to favour such an emergence, such as microbiological adaptation, globalisation of agriculture, food production and trade, human behavioural factors, and environmental changes. A systematic method for monitoring changes in these risk factors and in conditions associated with such outbreaks may increase alertness, resulting in improved surveillance. Besides, a careful review of past events could also help to identify key trends and provide guidance for the future[3].

Systematic monitoring of changes in risk factors associated with outbreaks would support prevention and control of pandemics.

Conclusion

Referring to the SARS epidemic in 2003, the WHO Director-General declared: 'We all swim in a single microbial sea.' Today, given the speed and volume of international traffic and trade, a new biological agent may spread around the world within a few hours and result in a global epidemic. This also affects the world of work, where the workplace may be, at the same time, a source of infection for workers and a bulwark against epidemics. Poultry workers, for instance, who, more than anyone else face intensive, daily occupational exposure to the risk of avian influenza are therefore one of the most likely vectors for a mutated H5N1 virus capable of human-to-human transmission. This means that the capacity to prevent global epidemics relies to some extent on the safe working conditions of workers. Conversely, because the epidemiological chain may start outside the confines of the workplace, there is a need to consider all collective responsibilities and means of control in order to identify and control the risk even before they enter into the workplace and pose an occupational risk to workers. In general, because the global threat of pandemics requires a global response, cooperation between various disciplines and authorities, as well as the systematic monitoring of changes in outbreak risk factors, are vital for effective forecasting, surveillance, prevention and control of emerging diseases and epidemics.

4.2.2. Workers' exposure to antimicrobial-resistant pathogens in the health care sector and livestock industry

Since their discovery during the 20th century, antimicrobial agents(29) have substantially reduced the threat posed by infectious diseases. However, this advantage is now seriously jeopardised by the emergence and worldwide spread of antimicrobial-resistant organisms, which make infections more difficult to treat, and increase the severity of illness and health care costs[78][79][80][81].

The overuse or misuse of antibiotics has led to drug-resistant organisms, endangering the effectiveness of antibiotics.

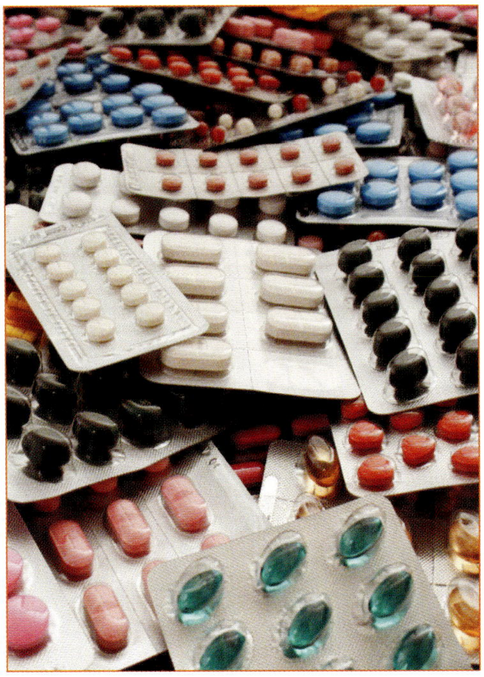

Antimicrobial-resistant organisms are bacteria and other organisms that have developed a resistance to certain antimicrobial agents. Drug-resistant organisms were first noted in the 1940s with penicillin resistance of Staphylococcus aureus[82]. The widespread use of antibiotics for human health, veterinary purposes and as animal growth promoters, as well as the natural evolution of bacteria, has resulted in the apparition of drug-resistant organisms[83]. Bacteria constantly adapt to their environment and have the ability to take on the characteristics of other bacteria. When antibiotics are used incorrectly — for instance, for too short a time or in too small dosage — the weak and susceptible bacteria are killed, while the more resistant ones survive and multiply[81]. Moreover, organisms that develop resistance to one antibiotic have the ability to develop resistance to other antibiotics. This is called cross-resistance[84].

Examples of drug-resistant organisms include[81][85]:

- MRSA: methicillin/oxacillin-resistant Staphylococcus aureus
- VRE: vancomycin-resistant Enterococci
- ESBL (extended-spectrum beta lactamases) producing bacteria
- PRSP: penicillin-resistant Streptococcus pneumoniae
- GISA: glycopeptide-intermediate Staphylococcus aureus
- VISA: vancomycin-intermediate Staphylococcus aureus
- MDR-TB: multidrug-resistant tuberculosis
- XDR-TB: extensively drug-resistant tuberculosis
- multi-resistant Escherichia coli and salmonellae
- carbapenem-resistant Acinetobacter
- ciprofloxacin-resistant Neisseria gonorrhoeae.

There is evidence that resistant organisms can move from human to human or from animals to humans by direct contact or inhalation, hence posing a health risk to health care workers, especially in hospitals, and to workers in contact with animals in

(29) Antimicrobials include antibiotics (i.e. naturally occuring chemicals), synthetic antibacterial agents, as well as compounds that affect other microorganisms, like parasites{88}.

veterinary services, the livestock industry and in the food-manufacturing industry. These resistant organisms may then colonise the workers exposed or transfer their resistance genes to bacteria that are endogenous to humans[86][87][88][89].

Especially intensive farming is a major source of overuse of antimicrobials as livestock is routinely fed with antibiotics to promote growth and to compensate for crowded, insanitary conditions conducive to infection[88][90]. In North America and Europe, an estimated 50% in tonnage of all antimicrobial production is used in food-producing animals and poultry[79]. In the United States, it has been estimated that the non-therapeutic use of antimicrobials in livestock production comprises 60-80% of the total antimicrobial production[90]. To overcome this problem, an EU-wide ban on the use of antibiotics for non-medical purposes, e.g. as growth promoters in animal feed, came into effect in January 2006. This is part of the Commission's strategy to combat the threat to human, animal and plant health posed by antimicrobial-resistant organisms due to the overuse or misuse of antibiotics[91][92].

An EU-wide ban on the use of antibiotics for non-medical purposes in farming came into effect in January 2006.

The continuous appearance of new antimicrobial-resistant organisms is inevitable, although controllable[85][93]. Vancomycin-resistant Staphylococcus aureus (VRSA) is not yet found in nature, but it is believed it will emerge or evolve from VISA[85]. In 2003, proliferation of new multi-drug resistant strains of the gram-negative E. coli was observed in the UK. These bacteria produce a new class of extended-spectrum beta lactamases (ESBL) called CTX-M. ESBL-producing bacteria have been known since the 1980s and are able to resist penicillins and cephalosporins, which are the most important classes of antibiotics and account for two thirds of all antibiotic use. Additionally, CTX-M producing bacteria have developed resistance to another important class of antibiotics, the fluoroquinolones. The common pathogen E. coli has hence evolved over a short period of time from one of the weaker into one of the more resistant members of the Enterobacteriaceae family. These shifts are occurring worldwide[81].

The inhalation of virginiamycin-resistant gram-positive bacteria found in the air of swine facilities could contribute to the appearance of quinupristin-dalfopristin-resistant gram-positive infections in humans, leaving few or no treatment options for the affected individual. Indeed, bacteria expressing resistance to virginiamycin are cross-resistant to quinupristin-dalfopristin, which is an injectable streptogramin A and B combination often used as a last resort for multidrug-resistant gram-positive infections characterised by MRSA and glycopeptide-resistant E. faecium and coagulase-negative staphylococci. Also, the presence of airborne clindamycin-resistant coagulase-negative staphylococci and viridans group streptococci raises the question as to whether these organisms could serve as reservoirs of clindamycin-resistant genes and of erythromycin-resistant genes[94], passing on clindamycin resistance to more pathogenic species as described above[93].

The emergence of drug-resistant tuberculosis poses a serious threat to public health. One third of the people in the world are infected with 'dormant' tuberculosis (TB) bacteria. The infected person may become ill with TB if the bacteria become active. This may happen if the person's immune system is weakened, for example as a result of HIV, advancing age or some medical conditions. Four standard or first-line drugs can be used to treat TB. However, a new multidrug-resistant TB (MDR-TB) has appeared. MDR-TB is resistant to the first-line drugs but can be treated with so-called second-line drugs. Nevertheless, second-line drugs have more side-effects, take longer to cure the illness and are more expensive. Furthermore, a new form of TB called extensively drug-resistant tuberculosis (XDR-TB) has emerged. XDR-TB is resistant even to the second-line drugs. In early 2006, a joint study by WHO and the

A new form of tuberculosis resistant to second-line drugs threatens public health also in Europe.

US Center for Disease Control and Prevention documented for the first time cases of TB that were extensively resistant to current drug treatments. XDR-TB was identified in all regions of the world, although it is still thought to be relatively uncommon. In September 2006, reports from KwaZulu-Natal province in South Africa of high mortality rates with XDR-TB in HIV-positive people increased the concerns about the emergence of XDR-TB. There is an urgent need to address XDR-TB, especially in areas of high HIV prevalence[95]. XDR-TB also represents a public health threat for Europe. Recent data has shown the presence of resistance to second-line drugs in most surveyed European countries. The strengthening of surveillance of resistance to first- and second-line drugs has become a priority in the context of TB control at regional level. Current initiatives for monitoring MDR-TB trends and burdens in Europe should be explored as potential opportunities for beginning or expanding the monitoring of second-line drug resistance[96]. Among the WHO recommendations on drug-resistant TB surveillance methods and laboratory capacity measures, one measure is aimed specifically at tackling infection control and protecting health care workers, especially in areas of high HIV infection[96].

Although many studies are available on antimicrobial resistance, the comparison of data is difficult due to unharmonised experimental conditions. In order to overcome this problem, the European Antimicrobial Resistance Surveillance System(30) (EARSS) started in 1999 to collect standardised data, focusing especially on gram-positive pathogens. It is a common view that resistance surveillance should focus mainly on MRSA and other gram-positive organisms. However, infections with gram-negative bacteria are quite frequent in Europe. In Estonia, for instance, high resistance and therapy failures are frequently associated with gram-negative bacteria. Since 2005, in addition to the information on gram-negative E. coli, EARSS has extended its data collection to gram-negative Pseudomonas aeruginosa and Klebsiella pneumoniae. Moreover, some findings show that different habits of antibiotic usage in European countries — such as the number of antibiotic prescriptions and the preference of different antibiotic groups observed between the Northern, Central and Eastern European countries — probably influence the spectrum and susceptibility pattern of invasive pathogens. Hence there is a need for international conventional surveillance systems to be modified according to local situations, and for the inclusion of additional diagnostic methods[97].

Routes of exposure and workers at risk

A key to understanding antibiotic resistance is acknowledging its inevitability. Four general mechanisms of resistance have been identified in bacteria: its ability 1) to reduce the intracellular concentration of the antibiotic; 2) to inactivate the antibiotic; 3) to modify the target site for the antibiotic; and 4) to eliminate the target site[85].

Drug-resistant organisms spread the same way other bacteria and organisms spread. They can spread from human to human or from animal to human[86]. Workers are at higher risk of infection if they[85]:

- have an existing severe illness;
- have an underlying disease or suffer certain health disorders such as chronic renal disease, insulin-dependant vascular disease, dermatitis or skin lesions, chronic respiratory system disease, etc.;
- have previously been exposed to antimicrobial agents;
- have undergone an invasive procedure such as dialysis and catheterisation;

(30) http://www.rivm.nl/earss/about/

- have previously been colonised by an antimicrobial-resistant organism;
- are elderly or on immune-suppressing medication.

Workers most at risk are the ones in contact with infected people — such as health care workers and especially hospital workers — but also laboratory workers and workers in contact with infected animals in the livestock and food industry (animal handlers, farmers, workers in broiler houses, workers in poultry production, slaughterers, butchers, etc.)[98][99][100].

In the health care sector, skin contact with devices or surfaces contaminated with body fluids from an infected person, or skin contact with an infected person, are the most common routes of exposure[85]. MRSA, for instance, is not airborne but usually spreads through physical contact[101].

Health care workers and those in contact with animals can be exposed to drug-resistant organisms.

Hospital workers are more likely to be exposed to drug-resistant organisms because of the number of patients with whom they come into contact in a single shift[82]. Hospitals are in any case a critical component of the antimicrobial resistance problem. Indeed, the combination of highly susceptible patients, intensive and prolonged antimicrobial use, and cross-infection may lead to nosocomial([31]) infections with highly resistant pathogens in patients, who then become a reservoir of contamination for staff. Failure to implement simple infection control practices before and after contact with patients, such as washing hands and changing gloves, is a common cause of cross-contamination between staff and patients[79][85].

Furthermore, the drug-resistant bacteria can be transmitted from animals to humans. In the livestock and food industry, a high density of animals, poor hygiene in the working environment and animal confinement areas, failure to control the risk of infection and to control the use of antibiotics, as well as stress reactions among the animals, are common causes of the promotion of drug-resistant pathogens, which can be passed to humans[100]. Resistant bacteria may colonise the human gastro-intestinal tract and transfer resistance genes to human endogenous flora[86]. Epidemiological studies have traced resistant human infections directly to specific livestock and poultry operations[90]. For instance, evidence of transfer of drug-resistant E. coli from pork and poultry was found in turkey farmers, broiler farmers, laying-hen farmers, turkey slaughterers, broiler slaughterers and pork slaughterers. Workers handling animal faeces are especially at risk[98][102]. Unlike many purely hospital pathogens, E. coli has a huge reservoir in the human intestinal tract where resistant strains can be carried for a long time, making the analysis complicated. With regard to the new apparition of CTX-M producing E. coli first noted in the UK (see above), more research is needed to identify all potential reservoirs and contamination routes, not only in the food industry but also in hospitals[81].

Airborne exposure is another pathway for the transfer of drug-resistant bacteria from animals to humans. Until now, little research has been conducted regarding airborne antibiotic-resistant bacteria within animal industrial facilities, but workers were found to be exposed to antibiotic-resistant pathogens via dust and bioaerosols in animal confinement facilities[99]; for example, in concentrated swine feeding facilities[93].

([31]) Nosocomial infection: infection acquired in an hospital that was not present or incubating prior to the patient being admitted to the hospital, but occurred within 72 hours after admittance to the hospital. Taken from: http://www.medterms.com/script/main/art.asp?articlekey=4590

Furthermore, there is evidence that activities such as waste collection, waste treatment and sewage treatment expose workers to bacteria and other pathogens (see '4.2.5 Biological risks in the management of solid waste'), and hence potentially to drug-resistant bacteria. Indeed, antimicrobials and antimicrobial-resistant pathogens are present in sewage from Confinement Animal Feeding Operations (CAFOs)[103], and in sewage and waste from pharmaceutical plants, hospitals, and also household waste, which contains microbes resistant to triclosan, quaternary ammonium compounds, alcohol and bleach found in toothpastes, kitchen plastics, cement and paints[104].

Infected or colonised workers may in turn become reservoirs of drug-resistant bacteria that can spread to the broader community[93][99].

Health outcomes generated

Drug-resistant organisms lead to more treatment failures and increased morbidity and mortality.

Resistant organisms may colonise the workers exposed or transfer their resistance genes to bacteria that are endogenous to humans[86][87][88][89]. In principle, colonisation rarely becomes an infection unless the bacteria are spread to a different and susceptible part of the body[85]. In that case, consequences include infections that would not have otherwise occurred, increased severity of infections and increased frequency of treatment failures, and even death in some cases. While it is difficult to quantify the total impact of antibiotic-resistant bacteria on health, it is clear that morbidity and mortality are increased by delays in administering effective treatment for resistant infections[105][106].

VRE is a major cause of post-surgical infections[104]. In the case of MRSA entering the human body, the same symptoms appear as with any other type of Staphylococcus aureus bacteria. The skin appears red and inflamed around wound sites. The subsequent health outcomes vary from minor infections, such as boils or pimples, to serious infections such as urinary tract infections including kidney failure, pneumonia, blood infections, toxic shock syndrome, and even death. Symptoms in serious cases may include fever, lethargy and headache[101].

E. coli bacteria normally colonise the intestinal tract without harm, but they are also a common cause of urinary tract infections such as cystitis, and of blood poisoning. Although many of the infections caused by CTX-M-producing E. coli are reported as common urinary tract infections, most of the affected patients have further complex health problems. In some cases, the infection has even led to death. In general, the infection was observed in elderly patients with underlying disease, a history of recent hospitalisation or antibiotic therapy[81].

The bacteria detected within the air of swine facilities are associated with diverse human infections[93]:

- Enterococcus, particularly E. faecalis and E. faecium, has emerged as one of the leading causes of nosocomial bacteremias, urinary tract infections, and wound infections in the United States;
- Coagulase-negative staphylococci are the third most common causes of nosocomial infections and the most common causes of nosocomial bacteremias.

Co-exposures to other aerosols and gases such as organic dusts, moulds and ammonia, for instance, in the swine environment, have been shown to induce symptoms associated with chronic bronchitis, including a persistent cough characterised by expectoration, which may increase the potential for spread of antibiotic-resistant organisms into the community[93][107].

Swine confinement facility

Most alarming of all are diseases where resistance is developing for virtually all currently available drugs. This raises the spectre of a post-antibiotic era. However, even if the pharmaceutical industry were to step up efforts to develop new replacement drugs immediately, current trends suggest that some diseases will have no effective therapies within the next few years[79].

Prevention

In the last 25 years, it has been possible to develop only one new class of antibiotics[90]. At present, the average cost of developing a new drug is around €500 million, and industrial incentives seem insufficient to overcome this barrier fast enough to secure continued access to effective drugs[108]. In view of the growing problem of antimicrobial resistance, urgent steps have to be taken to preserve the effectiveness of existing medicines for treating illness in both humans and animals. For instance, vaccination is an indirect but very powerful and cost-effective way to control infection that also reduces the need for antibiotics in the first place. Genomics offers a new and promising route to vaccine development. Exploitation of genomics is also likely to speed the arrival of new generations of cheaper drugs. For instance, the 'X-TB' project, a European project initiated under the fifth European Framework Programme (FP5) for Research and Technological Development (1999-2002), combines proteomics with structural and functional genomics approaches in order to develop new drug targets and new therapeutic compounds to treat tuberculosis[108].

In hospitals, antimicrobial-resistant nosocomial infections are expensive to control and extremely difficult to eradicate[79]. The basic measure against spreading antibiotic-resistant organisms and other pathogens responsible for epidemics such as SARS[109] among health care staff, is still thorough hand washing. Guidelines for the correct way to wash hands are available, for instance, from the US Centers for Disease Control and Prevention (CDC)[109]. However, broader measures, also at the collective and organisational levels, are necessary to ensure effective prevention of the spread of drug-resistant pathogens, and epidemics in general[85][110][111][112][113][114][115][116][117][118][119][120][121]. These include:

- improvement of work organisation (shift schedule, etc.)
- patient isolation

- restrictions on patient moves
- dedicating equipment such as stethoscopes, bedside commodes or thermometers, to one patient or one group of patients
- regular hospital cleaning
- aseptic techniques for patient-care equipment and the working environment
- use of safety-engineered devices (for example, retraction or shields for sharp instruments)
- appropriate handling and disposal of sharps (for example, needles) and clinical waste (waste generated during patient care)
- use of personal protective equipment, such as gloves, mask, goggles, gowns, plastic aprons
- changing gloves and washing hands, especially after contact with body fluids, and even between procedures on the same patient to prevent cross-contamination to different body sites
- training workers on correct hand washing, use of safety devices and safe disposal, etc.

Preserving the effectiveness of drugs requires cooperation between various actors.

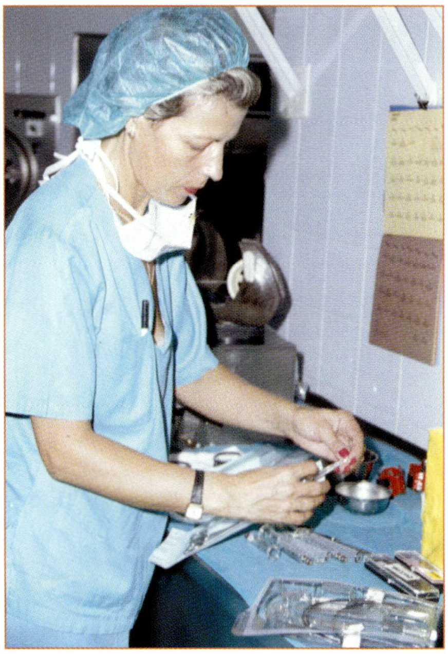

Handing sharps in hospitals — Instituto Nacional de Seguridad e Higiene en el Trabajo, Spain.

Further recommendations to limit antibiotic resistance and preserve the effectiveness of existing medicines overlap with the public health sphere and advocate the strict control of antibiotic use and elimination of misuse and overuse. In hospitals, control programmes and committees in charge of overseeing antibiotic use should be established, and guidelines for correct antibiotic treatments developed. Also, hospitals should control and monitor pharmaceutical companies' promotional activities within the hospital environment and ensure that such activities have educational benefit[83][105][114][116].

The EU-wide ban on the use of antibiotics for non-medical purposes, e.g. as growth promoters in animal feed, mentioned earlier has an obvious impact on the food industry[91][92]. There has been considerable international concern that banning the use of growth promoters would have a wide range of negative effects, not only regarding the efficiency of meat production, but also with regard to animal health and food safety. However, the Danish experience — Denmark had already opted for a complete ban of antimicrobial growth promoters in 1998-1999 — shows that the use of antimicrobials for the sole purpose of growth promotion can be discontinued without engendering any serious negative effects[122][123]. However, it should be noted that since the ban, the use of antibiotics for veterinary purposes, i.e. to control diseases, has increased by 50% in Denmark, partly to control infections that might otherwise have been suppressed by growth promoters[124].

Beside the use of antibiotics as growth promoters, their use for veterinary purposes should also be strictly controlled by:

- using agents with limited spectra of activity and
- developing guidelines for the proper use of antibiotics in animals[89][90].

Methods should also be developed to monitor routinely the respiratory and immune system status of workers, especially in confinement livestock operations, and to check whether they are colonised by antibiotic-resistant pathogens[99]. Additionally, better animal housing and better animal hygiene should be sought to avoid animal diseases and reduce the amount of antibiotics used in farming[89][124]. Proposed solutions range from the introduction of automated air quality monitoring systems in confinement livestock operations to a shift away from intensive farming, in which animals are kept in close proximity in often insanitary and stressful conditions[90].

The fifth European Framework Programme (FP5) for Research and Technological Development (1999-2002) devoted more than 50 million EUR to projects tackling the problem of antimicrobial resistance. The 'ARPAC' (32) project, for instance, is gathering data on antibiotic consumption and pathogens' resistance with the aim of developing harmonised strategies for prevention and control of antibiotic resistance in European hospitals. Indeed, antibiotic consumption must be monitored and linked to both resistance data and clinical outcomes. In this regard, 'DEAR' (33) addresses the dynamics of the evolution of antimicrobial-drug resistance, for instance the effects of different antibiotic dosage regimes. Additionally, modern DNA technology is essential to accelerate the development of new, quick and inexpensive diagnostic tests. Such tests will enable physicians to identify pathogens and their resistance properties, which is a key element for the prudent prescription of appropriate antibiotics. In this regard, one of the aims of the 'Pseudomonas Virulence'(34) project is to develop a new, DNA chip technology-based diagnostic test for pseudomonas aeruginosa — a common and particularly dangerous cause of nosocomial infections. The results of this project may be applicable to many other human pathogens. 'Dissarm' (35) is another project aimed to develop rapid and highly sensitive diagnostic tests for multi-drug resistant strains of tuberculosis.

In conclusion, it is inevitable that antimicrobial-resistant organisms will continually evolve[85][93]. The challenge is to identify them rapidly as they emerge, assess their potential impact on health, understand their multiple genera in different environments as sources of human exposure, measure their prevalence in hospitals and communities, and devise policies and procedures to minimise their spread[81][93]. For this to be successful, cooperation between various agents is required, including the OSH, public health, animal health, farming and food industries, environmental services, and even the social and health economy because antimicrobial resistance addresses wide-ranging socio-economic and political issues, such as physicians' prescribing behaviour, medical reimbursement and public expectations[108].

The emergence of drug-resistant organisms is inevitable but controllable.

4.2.3. Occupational exposure to endotoxins

Endotoxins are toxins built of polysaccharide and phospholipid substances that are integral parts of the outer cell wall of bacteria. In general, the term 'endotoxin' is used

(32) http://cordis.europa.eu/data/PROJ_FP5/ACTIONeqDndSESSIONeq112482005919ndDOCeq758ndTBLeqEN_PROJ.htm

(33) http://cordis.europa.eu/data/PROJ_LIFE/ACTIONeqDndSESSIONeq703200595ndDOCeq66ndTBLeqEN_PROJ.htm

(34) http://cordis.europa.eu/data/PROJ_LIFE/ACTIONeqDndSESSIONeq703200595ndDOCeq250ndTBLeqEN_PROJ.htm

(35) http://cordis.europa.eu/data/PROJ_LIFE/ACTIONeqDndSESSIONeq703200595ndDOCeq69ndTBLeqEN_PROJ.htm

to refer to lipopolysaccharide (LPS) of the outer membrane of gram-negative bacteria — LPS is composed of a lipid A, a core polysaccharide and an O-antigen (O polysaccharide side chain). When the bacterial cell grows or is destroyed, its endotoxins are released. Endotoxins are responsible for many of the virulent effects of gram-negative bacteria[125].

Endotoxins are found in occupational settings where organic dust is present.

In the environment, endotoxins are mostly found in organic dust — containing particles of plant, animal or of microbial origin — which is widespread in occupational settings. As bacterial flora differs between work environments, different patterns of exposure to endotoxins can be found even in similar workplaces.

Worker groups most at risk and workplaces concerned

Endotoxins can be found at high concentrations in all occupational environments where organic dust is present. What was initially considered to be a problem only for a few industries and activities, such as cotton farming, or handling of mouldy hay by farmers, turned out to also affect people working in swine confinement facilities, chicken housing, handling garbage and sewage water, and even at indoor workplaces suffering mould growth[126].

Delivery of unseparated domestic waste in a sorting plant. Photo by Sirpa Laitinen, © FIOH

Of all the workers exposed to bioaerosols, farmers are likely to be at highest risk. Farming covers a large variety of tasks, techniques and products. Bacteria can originate from manure, skin scales, straw, hay and grain etc.[127]. The type, duration, and level of exposure to endotoxins varies greatly between grain production and greenhouse farming, and between dairy farming and poultry production. These working environments differ considerably in terms of concentrations of endotoxins in ambient air, ranging from an average of 3 endotoxin units per cubic meter (EU/m^3) in greenhouses to 610 EU/m^3 in swine confinement houses[128].

Recent research suggests that other groups of workers — such as office workers[164], and animal technicians, as well as scientists working with rodents — are also at increased risk of exposure to endotoxins[165].

Sector	Workplaces	Workers most at risk	References
Agriculture and forestry	Agricultural farms	Farmers and their families Agricultural workers	[129][130][131][132] [133][134]
	Crop growing	Crop farmers Hop growers	[132][135]
	Animal houses	Animal farmers Swine husbandry Workers in poultry houses	[136][137][138]
	Woodworking shops	Joinery workers	[139]
Textile industry	Textile mills	Jute mill workers Cotton textile workers Cotton spinning mill workers Textile plant workers Nylon plant workers	[140][141][142][143] [144][145][146][147] [148]
Sewage, waste and recycling	Sewage treatment	Sewage treatment plant	[148][149][150]
	Solid waste	Household waste collectors Workers in recycling, handling and sorting	[151][152][153][154]
Paper industry	Paper mill Factories producing soft tissue paper	Paper mill workers Paper production unit workers Workers in pulp and paper mills Workers in paper and paper board mills	[155][156][157][158]
Metal industry		Metalworking	[159][160][161][162]
Food production and processing	Seafood industry	Workers in seafood industry plants	[163]

Toxicity mechanisms, routes of exposure and health effects

In vivo, gram-negative bacteria are most likely release minute amounts of endotoxins when growing. Furthermore, it is known that small amounts of endotoxin may be released in soluble form, especially by young cultures. However, for the most part, endotoxins remain associated with the cell wall until disintegration of the bacteria, i.e. autolysis of the bacteria, external lysis mediated by complement and lysozyme, or phagocytic digestion of bacterial cells[125].

Endotoxins are responsible for many of the virulent effects of gram-negative bacteria.

The biological activity of endotoxins is associated with the lipopolysaccharide (LPS), which is the best-studied pyrogen[166]. Toxicity is associated with the lipid component (lipid A), and the immunogenicity with the polysaccharide components. Both act as determinants of the virulence of gram-negative bacteria[167][168]. The immune stimulatory capacity of endotoxins can only be inactivated at high temperatures (for example, 160 °C for four hours). Therefore, endotoxins are active for much longer than the lifetime of the bacteria themselves[128].

LPS elicits a variety of inflammatory responses in animals and humans. Indeed, it plays an important role as a surface structure in the interaction of the pathogen with its host. For example, LPS may be involved in adherence (colonisation), or resistance to phagocytosis, or antigenic shifts that determine the course and outcome of an infection[125][169]. Exposure of experimental animals to living or dead gram-negative bacteria, or purified LPS, via intraperitoneal or intravenous route causes a wide spectrum of nonspecific pathophysiological reactions — such as fever, changes in white blood cell count, disseminated intravascular coagulation, hypotension and shock — and results in death in most mammals, even for intraperitoneal or intravenous injections of low doses of endotoxins. The sequence of events follows a regular pattern: (1) latent period; (2) physiological distress (diarrhoea, prostration, and shock); (3) death. How soon death occurs varies, depending on the dose of the endotoxin, the route of administration, and the animal species. Animals vary in their susceptibility to endotoxins[125].

Research in recent years has revealed major clinical effects caused by exposure to organic dust and identified some causative agents such as bacterial endotoxins, moulds, and various allergens[126]. In humans, endotoxins have been implicated as the aetiological agent of a variety of pathologies. These range from fever and a variety of health effects with a major public health impact — such as infectious diseases, acute toxic effects, allergies, organic dust toxic syndrome (ODTS), chronic bronchitis, and asthma-like syndromes[128] — to lethal effects such as septic shock, organ failure and death[170]. Response to endotoxin exposure in humans varies as a function of dose, route of exposure, and rapidity of release into the blood circulation. Exposure to sub-lethal doses of endotoxins causes dramatic changes in human body temperature, in the haematological, immune, and endocrine systems, and in metabolism[171]. Short-term exposure to endotoxins in the air at levels above 45 EU/m^3 may be linked with decreases in lung function over the course of a single day. Longer-term exposure to endotoxin levels as low as 10-28 EU/m^3 may be linked with chronic diseases in lung function[172].

Regarding inflammations of the respiratory tract, the response is mediated by a number of bioactive molecules mainly secreted by activated phagocitic cells, for the most part macrophages, and include among others proinflammatory cytokines IL-1, IL-6, IL-8, and TNF-a which give rise to general and non-specific symptoms such as fever, headache, malaise, respiratory symptoms (tightness of chest, dry cough), and lung function decrements[173].

However, a complex, dose-dependent, non-linear relationship between environmental exposure to endotoxins and the outcome of immune responses has been revealed in several experiments. Although seemingly paradoxical, endotoxins' dual nature ultimately may serve to improve our understanding of how such bioactive agents can interact with and guide our immune systems in both health and disease[128][174][175][176]. Indeed, it has been found that exposure to endotoxins — as well as fungal spores and other pathogen-associated molecular patterns (PAMPs) — may induce, but conversely may also protect from, asthma, atopy, respiratory allergies and sensitisation to allergens[128][177][178][179]. For instance, the role of LPS in the regulation of Th2 ([36]) responses and in immunoglobulin E (IgE) production is apparently conflicting, as both protection and exacerbation have been reported[168]. The existence of different asthma phenotypes may account for the opposite response patterns regarding asthma[128].

Additionally, in farm environments for instance, microbial exposure including endotoxins may have protective effects that probably develop during childhood and can still be observed at an adult age. It even seems that exposure at an adult age may stimulate the innate immune system and have a protective effect against developing

([36]) Th cell: 'lymphocytes [...] that bear antigen receptors on their cell surface to allow recognition of foreign pathogens.'

Th2 cell: 'Th cell of type 2, that are essentially anti-inflammatory but that promote allergic responses. [...] Many researchers regard allergies as Th2 weighted imbalance.' Taken from: Berger, A., 'Th1 and Th2: What are they?'. *BMJ*, Vol. 321, No 7258, 12 August 2000, p. 424, http://bmj.bmjjournals.com/cgi/content/full/321/7258/424

allergies. However, it is hypothesised that reversal of atopy might occur as a result of high exposure to endotoxins and other PAMPs[178].

Farming

Dust endotoxin also appears to serve as a marker for other innate immune-stimulatory microbial components PAMPs, such as bacterial DNA, which may augment and steer endotoxin-initiated immune responses in an immune-regulatory direction. These findings support the premise that the differences in health outcomes from endotoxin exposure are due to important moderating variables, such as age of exposure, timing of exposure relative to disease development, dose and frequency of exposure, co-exposures, and genetic predispositions in response to endotoxin[179].

Last but not least, there is increasing evidence that the interactions between genes and the environment may play a critical role in the pathogenesis of complex diseases exhibiting a heritable component, such as asthma. Indeed, recent experiments support the existence of an 'environmental switch' capable of eliciting different and even opposite immune phenotypes. These gene-environment interactions might affect the emergence of disease differentially depending on the interplay between environmental exposure and genetic background of individuals. A functional implication of the 'endotoxin switch' is that endotoxin exposure may result in different responses in humans depending on the environmental context[167][168].

Respiratory symptoms[132], such as ODTS[131][134][180] or allergic and non-allergic asthma, atopy[181] and eye symptoms[133][164][182], appear to be common among farm workers. Fever, chills and respiratory symptoms, such as chest tightness, cough and shortness of breath, were observed in sugar beet processing plants[183]. An increased exposure to bacterial endotoxin in airborne dust related to byssinotic symptoms was found among jute mill workers[141]. Similar findings were reported for workers in the cotton, flax, and mushroom industries[140][143][184][185][186]. An increased prevalence of mucosal and skin symptoms[151][187], work-related ODTS, chronic bronchitis, and respiratory symptoms have been observed among waste handlers[152][188][189][190].

Measurement methods and risk assessment

Endotoxins are usually measured in samples of airborne and settled dust. Soluble endotoxins are determined by the Limulus amebocyte lysate (LAL) method, which has been widely used for about 25 years. Bacterial species and strains seem to be an important factor for the solubility of endotoxin. However, it is suggested that non-soluble endotoxins should also be quantified in the assessment of health risks of endotoxins to humans[127]. This, together with the problems of variations in sensitivity and specificity of LAL to endotoxin, and of the limited supply of limulus (horseshoe crabs) has opened a new era in endotoxin testing[166].

Owing to different measurement protocols, large inter-laboratory variations in the results of analyses of endotoxins can be found. Therefore, several research articles support the need for standardisation of methods for endotoxins measurements, even through the introduction of a new European standard, to allow for acceptable inter-laboratory precision and accuracy[128][152][191][192]. Risk assessment is seriously hampered by the lack of valid quantitative exposure assessment methods[192][193][194].

There is a need for standardised methods for endotoxins measurements.

Occupational exposure limits (OELs) for bacterial endotoxins have been proposed and discussed[195][196], but there are no common regulatory standards. The Dutch Expert Committee on Occupational Standards recommends a health-based occupational exposure limit for airborne endotoxin of 50 EU/m^3 (approximately 4.5 ng/m^3) based on personal inhalable dust exposure, measured as eight-hour time weighted average[197]. The Swiss indicative limit value is set at 1000 EU/m^3[198].

In Germany, national legislation (Biological Agents Ordinance — BioStoffV and Technical Rules on Biological Agents, TRBA) and the recommendations of workers' compensation boards define standardised methods for the assessment of airborne mould, bacteria, and endotoxins. Policies and practices are available for the measurement of airborne bioaerosols and for the interpretation of measurements relative to the standards. For example, in the agriculture sector, these standardised measurement procedures have proven suitable for use in livestock buildings and confirmed the often high concentrations of airborne biological hazards in the agricultural sector reported in the literature[199].

In Poland, the Institute of Rural Medicine in Lublin has drafted proposals for threshold limit values of occupational exposure to bioaerosols associated with plant and animal dusts, including bacterial endotoxins, gram-negative bacteria, mesophilic bacteria, thermophilic actinomycetes and fungi. These proposals could be considered to be a starting point for developing appropriate facultative standards that would facilitate the practical implementation of Directive 2000/54/EC on the protection of workers from risks related to exposure to biological agents at work (37). In the meantime, it is essential to be strict in following the binding concentration limits of plant and animal dusts in the air[194].

Possible prevention measures

Endotoxins are heat stable and are not destabilised when boiled for 30 minutes. However, certain powerful oxidising agents such as superoxide, peroxide and hypochlorite degrade them[125].

The toxic portion of LPS (the lipid A portion) is relatively similar across a wide variety of pathogenic strains of bacteria, making this molecule an attractive target for the development of an LPS antagonist. Research focused on the design of various lipids A analogues has led to the development of E5564, which has proven to be an advanced, unique and highly potent in vivo antagonist of endotoxins and may hence be of benefit in a variety of endotoxin-mediated diseases[200].

The following measures are recommended to minimise exposure to organic dusts[201]:

- operate within a controlled environment (cab, control room, etc.)
- ventilate confined or dusty areas using fans, exhaust blowers, filters, etc.
- move work outside when possible
- wear respirators, masks, or other protective equipment, especially in the agricultural sector
- use automatic feed-handling systems
- wet the top of silo before uncapping ensiled material

(37) 'Directive 2000/54/EC of the European Parliament and of the Council of 18 September 2000 on the protection of workers from risks related to exposure to biological agents at work (seventh individual directive within the meaning of Article 16(1) of Directive 89/391/EEC)', *Official Journal* L 262, 17 October 2000, pp. 21-45, http://europa.eu.int/eur-lex/lex/LexUriServ/LexUriServ.do?uri=CELEX:32000L0054:EN:HTML

- use wetting techniques when cleaning out grain bins or other dusty areas
- install covers over grain bins inside buildings to reduce dust production.

The efficiency of different types of respirators has been assessed to protect workers from exposure to bacterial aerosols, which includes endotoxins[202][203]. The latest testing indicates that paper masks are not suited for protection against bioaerosols, but that multi-layer filters in mouth-nose protectors (MNP) and filter face pieces (FFP)[72] are effective. A NIOSH-approved N95 respirator (38) called 'Tbc-mask' showed the highest efficiency, corresponding to category FFP2 (39) [72]. However, personal protective equipment should be used as the last possible prevention measure only when eliminating or reducing the level of risk to an acceptable level is not possible.

4.2.4. Moulds in indoor workplaces

Airborne moulds are ubiquitous in the indoor environment.

Indoor exposure to moulds and subsequent health issues have only been given close attention relatively recently[204][205]. To date, more than 100,000 species of mould have been identified, but it is estimated that there may be more than 1.5 million species worldwide. Mould organisms grow by degrading nutrients from organic substrates such as wood and wood products, fabrics, foodstuffs, plants and plant debris, and soil[206]. Airborne moulds are ubiquitous in the indoor environment, the most common ones being Cladosporium, Alternaria, Penicillium, Aspergillus, Mucor, Aureobasidium and Phoma species[205][207][208][209][210]. Several methods for the measurement of mould fungus concentration in workplace atmospheres have been described[211]. The application of polymerase chain reaction (PCR) and probe hybridisation techniques in detection of airborne fungal spores in environmental samples is still in statu nascendi[212][213][214][215].

Health-based exposure limits to airborne mould could not yet be established.

How to determine the actual extent to which exposure to moulds poses a health risk is still subject to debate[206][216][217][218][219]. Attempts have been made to identify the fungi responsible for specific symptoms, whether inflammatory or mycotoxic, attributed to mould exposure. However, the diversity of biological agents potentially present in the indoor occupational environment, and their various health effects on individuals, make it difficult to establish safe or unsafe levels of airborne mould concentrations. In addition, the interpretation of fungi concentration data remains difficult, first because mould concentration varies seasonally, geographically and according to the diurnal cycle, and secondly because the sampling and measurement methods are not standardised. Indeed, air-sampling methods currently used to estimate bioaerosol concentrations have not been demonstrated to accurately predict mould growth in a building[220]. Last but not least, there is a lack of epidemiological and clinical data that establish exposure-disease and dose-response relationships. Therefore, health-based exposure limits cannot yet be proposed[205][221][222][223].

Despite the lack of occupational exposure limits (OELs), various organisations have developed recommendations for addressing indoor fungal concentrations. However, there is little consistency among these recommendations.

(38) An N95 respirator is one of nine types of disposable particulate respirators. It filters at least 95 % of airborne particles, but it is not resistant to oil. Taken from: http://www.cdc.gov/niosh/npptl/topics/respirators/disp_part/default.html

(39) The classification of available filtering half masks is carried out according to European Norms (EN 149) (Filtering Face Piece = FFP available in three classes P1, P2 and P3 providing differing protection factors (levels) (efficiency: low, med, high).

While guidelines of the American Occupational Safety and Health Administration and Swiss indicative OELs[198] suggest that levels greater than 1,000 CFU/m³ (Colony-Formit Unit per Cubic Meter) constitute a probable contamination source, the European Commission establishes that levels above 500 CFU/m³ are an intermediate, and levels above 2,000 CFU/m³ a high source of contamination in indoor non-industrial workplaces. Investigation and remediation are considered to be required when indoor airborne fungal concentrations exceed 500 CFU/m³ and when occupants complain of non-specific health symptoms (for example, headaches, fatigue and coughs). However, levels in excess of these guidelines do not necessarily imply that the conditions are unsafe or hazardous. It was even observed that a large proportion of 'non-complaint' buildings — i.e. in which occupants do not have health concerns associated with the quality of the indoor air — show indoor ambient air fungal concentrations above 500 CFU/m³, and often levels even higher than the ones detected in buildings with complaints of non-specific health symptoms. Therefore, in addition to determining the number of airborne CFU/m³, the identification of predominant taxa, or at least fungi, is recommended to evaluate properly the hazard to workers. Moreover, as fungi concentrations vary between geographic location, and even within the same geographic location depending on the season, comparison of levels in non-complaint and complaint buildings collected at the same time is required for a scientifically sound evaluation of indoor fungal concentrations in complaint structures[205][221][224][225].

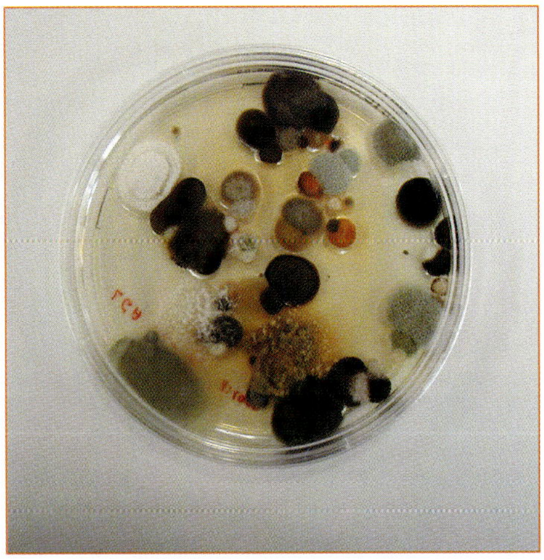

Moulds mixed culture — Berufsgenossenschaftliches Institut für Arbeitsschutz, Germany

The concerntration of airborne mould, and also the predominant type of fungi present need to be determined.

Mould growth and workers at risk

As moulds have no chlorophyll, they are obligate or facultative saprobes or parasites (40), meaning that they depend on an external source of organic material for growth. They reproduce typically by spores, which are small propagating structures that can produce a new individual, via sexual or asexual mechanism. Spores differ in number of cells, size (from 2 to 100 ìm), shape and colour. Most spores are adapted for airborne dispersal and can be introduced into the indoor environment through natural ventilation (open windows and doors) or mechanical ventilation, and can also be dispersed by insects, water, animals and humans[205][224][226][227][228][229]. Spores present in the building structure and materials remain in a dormant state unless the materials become humid.

(40) Saprobe: Fungus that cannot produce its own food (called heterotroph) but derives its food from *non-living* organic carbon sources. Parasite: Heterotroph that derives its food from the *living* cells of another living organism.

Facultative Parasite: Heterotroph that is primarily a saprobe, but when opportunity presents itself, can be a parasite.

Facultative Saprobe: Heterotroph that is primarily a parasite, but when opportunity presents itself, can become a saprobe.

Taken from: University of Hawai at Manoa, http://www.botany.hawaii.edu/faculty/wong/BOT135/Lect03_a.htm

However, even in the dormant state, they can be liberated into the indoor air and behave as organic dust, i.e. can become sediment on surfaces, be inhaled by occupants and deposit on the mucosal surface of the upper airways, or deposit on the eyes[207][230].

Many construction materials contain enough organic material to cultivate mould when wet[223]. Different types of moulds grow on different types of building materials, depending both on the nutrients and the amount of water present[230]. Moulds can grow in building structures (porous building materials), finishing materials and furnishings of the building (gypsum wallboard, carpets, flooring, etc.) and components of HVAC (heating, ventilation and air conditioning) systems[227][228]. Stachybotrys, for example, is frequently found on wet paper used in gypsum wallboard and other materials with high cellulose content[223].

Fungal indoor concentrations are influenced by several factors including:

- temperature
- high indoor humidity (more than 60%)
- condensation on windows and cold surfaces
- inadequate ventilation
- improper maintenance and equipment operations
- presence of some specific reservoirs of contamination, including plants and pets, and
- water intrusion into building structures; for instance, because of compromised building envelope (broken vapour barrier, air or moisture infiltration), inadequate insulation or sealants, leaky foundation and poor drainage, roof and plumbing leaks, water damage due to fire suppression efforts, etc.

Workers may be at risk of exposure to moulds in any indoor workplace.

Many mould-contaminated buildings suffer from chronic leaking through exterior wall and roof systems, sometimes as a result of poor consideration at the design stage of the possibility of rainwater penetrating the exterior surface once the building is finished[205][208][210][223][230][231][232].

The accumulation of dust, together with high humidity in components of HVAC systems, is a main source of microorganisms. Portions of ventilation systems near cooling coils and drains pans are prone to be exposed to high moisture levels for extended periods, and fibrous duct insulation materials are known to be sources of microbial contamination. Therefore, HVAC systems should be properly maintained, filters frequently replaced and ducts effectively cleaned[227][228]. The whole building in general should be kept clean and well maintained[205][208][210][231][232].

Workers may be at risk of exposure to moulds in any indoor workplace; for example, in offices, schools, hospitals, homes, and any commercial or residential buildings[219][233].

Higher risk of mould exposure is also found in solid waste or wastewater treatment, in cotton mills and in the agricultural sector in any activity involving exposure to mouldy hay, straw, or grain dust, for example, in grain storage areas, when unloading or uncapping silos, in cattle husbandry, etc. Agricultural workers most often suffer from mould-related diseases in winter and early spring, because the moulds have had time to develop in closed storage areas[209][217][218][222][228][234][235]. Hazardous materials removal workers, and in some cases construction workers, are also exposed to mould when performing mould remediation activities. This is a new and growing part of their work[236].

Health outcomes

Humans are at risk of indoor mould exposure when fungal spores, fragments or metabolites are released into the air and inhaled or physically contacted through dermal exposure[223].

A number of health problems related to exposure to indoor moulds have been documented. Both high-level, short-term exposures and lower-level, long-term exposures can result in ill-health[206]. The most common symptoms induced by exposure to indoor mould or mould spores are[207][209][222][223][224][226][229][230][231][237][238]:

- sick building syndrome (SBS)
- asthma, or exacerbation of asthma in mould-sensitive asthmatics
- allergic diseases
- increased rates of upper respiratory disease
- infection — people with suppressed immune systems are especially susceptible to fungal infections
- nose, throat or eye irritation
- runny nose, cough, congestion, headache and flu-like symptoms and
- skin irritation.

Allergies are probably the most common response to mould exposure. Atopic workers may develop allergy symptoms following skin exposure or inhalation of mould or mould products to which they have become sensitised. The fundamental understanding and close dialogue between workers, their management, health and safety officers, architects, engineers and building health specialists is essential in order to identify, evaluate, monitor and remedy building-related allergic reactions[207][239][240].

As part of their normal metabolism, moulds can produce a variety of volatile organic compounds (VOCs), which contribute to the musty, mouldy or earthy odours. Fungal VOCs may have irritant effects and provoke responses such as tingling and burning sensation of the skin, conjunctivitis, rhinitis and asthma. Airway inflammation and irritation of the mucous membrane can also be caused by airborne glucans from fungal cell wall, or by mechanical effect of spores and mycelia debris[222][237][240].

The concentration level at which indoor airborne mould represents a threat to health is still unclear.

Additionally, some moulds produce antibiotics and mycotoxins as by-products of their metabolism. Mycotoxins are typically cytotoxic, disrupting cell membranes and interfering with protein and RNA/DNA synthesis. However, specific human toxicity due to inhaled mycotoxins is not understood well and the likelihood that sufficient mycotoxins are airborne despite visible indoor mould remains unproven and controversial[222][226][241].

Finally, the level at which indoor airborne mould concentration becomes a threat to health is still unclear. First, the diversity of biological agents to which workers are often simultaneously exposed makes it difficult to determine which agent has which health effect and at which exposure level, all the more so because of the lack of known biological markers of exposure to fungi. In addition, the interpretation of mould concentration data remains complex, as mould concentrations vary seasonally, geographically and according to diurnal cycle, and as there are no standardised sampling and measurement methods. Additionally, not all persons exposed to mould will necessarily exhibit adverse health effects. Indeed, the susceptibility to exposure varies with the individual's genetic predisposition, age, state of health, and with concurrent exposures. However, there is general agreement that active mould growth in indoor environments is insanitary and that indoor mould must be removed[205][221][222][223][226].

Prevention

Policies and guidelines are available and recommendations have been formulated to minimise mould exposure and eliminate indoor moulds[223][226]. Multi-disciplinary input from the persons involved in building construction, services and controls, design, use and maintenance of buildings is required. The following measures should be considered to minimise the potential for mould growth[207][208][223][240]:

- minimising the exposure of interior building products to exterior conditions
- maintaining the integrity of the building impermeable envelope components through ongoing monitoring and inspections
- checking all material deliveries to validate that components are dry and clean, and rejecting wet or mouldy materials
- protecting stored materials from moisture
- preventing spillage of water within the building
- minimising moisture accumulation within the building
- achieving balanced control of thermal comfort and relative humidity in the building
- monitoring and maintaining installations, including heating, ventilation and air-conditioning (HVAC) systems, to ensure they remain clean and dry
- using filtration systems to prevent the ingress of mould spore from outdoor sources.

Moisture promotes mould growth and should be controlled.

Maintenance of air-conditioning system

Building health and any indoor pollutant problems should be addressed before the construction phase of any structure starts. An early, clear planning and understanding of the building's use should guide the careful selection of building systems and techniques, and furnishings and operating equipment. Although it is not possible to eliminate mould spores and nutrients completely from the construction process, it is possible to control moisture, which is one of the factors promoting mould growth. In this respect, construction managers should give careful consideration to the timing and scheduling of the project in order to avoid construction during damp or rainy seasons. If there is a risk of exposing materials to moisture during construction, appropriate 'mould-resistant' materials should be chosen to reduce the risk of mould growth. It should be ensured that concrete walls, beams and floors, wooden structural components, gypsum mouldings, and other materials are allowed to dry completely without being covered. Designers should provide proper system design and material selection to prevent water intrusion or condensation, but also to minimise microbial habitats and to avoid sources of particulates and other pollutants such as VOCs. Building operators must establish detailed guidelines on maintenance and inspection for the prevention and early detection of mould[207][223][228].

Proper design principles can reduce the risk of the HVAC system contributing to mould growth in a building. Good practices are available for ductwork design, cooling for dehumidification, and proper installation of humidification systems to reduce moisture in ductwork and the likelihood of mould growth[223].

Mould growth may be not visible but hidden in the walls or ductwork.

With regard to mould assessment in existing buildings, the thorough inspection of the premises and HVAC systems is a fundamental element. Checklists are available for the visual determination of mould. Even if the building inspection reveals no obvious mould growth, there may still be hidden growth within the walls, ductwork, or any other hidden locations. Therefore, where risk factors for possible mould growth are present — for example, history of water damage or building envelope failure, surface staining or mouldy odours — an intrusive inspection is necessary to determine the full extent of contamination. Intrusive inspections may involve peeling back areas of baseboard or vinyl wallpaper, removing sections of carpet or ceiling tiles, cutting holes into wall or ceiling cavities, or inspecting the HVAC systems, components and ductwork. Surface and air samples should be obtained for laboratory analysis[223].

In the case of fungal contamination of a building, the primary response must be prompt remediation of contaminated material and infrastructure repair. Certain precautions should apply to the handling, disposal, recycling, and transportation of mouldy materials. Training workers in this respect helps to minimise both the risk of exposure and cross-contamination during demolition and the handling of mouldy materials[223]. Remediation workers should wear gloves, eye protection, appropriate respiratory protection (at least a NIOSH-approved N95 disposable mask ([41])) and even full-face respirator, as well as clothing with head and foot covering for larger contaminated areas[223].

There is a need for criteria and measurement methods enabling to determine whether remediation has been successful.

Once the remediation of the mould has been completed, follow-up testing of the area is necessary to ensure the job was properly done and that the conditions for mould growth have been eliminated. This will help to prevent mould recurrence. Indeed, in some cases, the building was found to be still contaminated after remediation, or there was a recurrence of mould growth[242]. However, final clearance sampling practices for mould abatement projects have not been evaluated[204], and the use of air samples to evaluate the efficacy of a mould remediation may result in misleading conclusions and a false sense of security to building occupants[220]. Furthermore, the method of comparing indoor to outdoor mould levels — where a ratio of indoor to outdoor mould spore concentrations inferior to one would indicate the absence of mould growth and hence remediation efficacy — is controversial as the species of mould present indoors differ from the ones outdoors and hence has different effects on health at different levels[220]. There is a need for criteria and measurement methods enabling to determine whether remediation has been successful[204][220][242].

In conclusion, mould can potentially affect anyone in an indoor environment and is an important health issue. More research and systematic field investigation are still needed to provide harmonised measurement methods to assess better the risk of exposure, and to develop a better understanding of the health implications of indoor mould exposure[205][221][222][223][226].

4.2.5. Biological risks in the management of solid waste([42])

In the 1990s, several governments adopted new waste management policies with the primary aim of increasing the quantity of waste recycled. Under the EU Landfill

([41]) An N95 respirator is one of nine types of disposable particulate respirators. It filters at least 95 % of airborne particles, but it is not resistant to oil. Taken from: http://www.cdc.gov/niosh/npptl/topics/respirators/disp_part/default.html

([42]) Activities related to sewage treatment and soil decontamination are not dealth with in this review. [CV0]Spacing cannot be arranged without touching the notes the author instructed to leave alone.

Directive[243], the next 20 years should see a dramatic decrease in the amount of biodegradable municipal waste sent to landfill[244]. As a consequence, the recycling industry is a relatively new but expanding business: the number of workers involved in waste treatment has been increasing and will rise steadily[244][245].

More and more workers are exposed to biological agents in the waste treatment industry.

Because of the lack of statistics available on this sector, it is difficult to describe it in terms of numbers of workers and companies, and specific indicators for occupational accidents and diseases. In France, it is estimated — probably under-estimated — that around 100,000 workers are employed in a sector related to waste management, with about half of these involved in the collection and treatment of domestic waste[246]. In the UK, it is estimated that around 160,000 workers are employed in the waste industry, but many more are employed in other activities associated with specific recyclables and ancillary activities such as transportation. Recent research estimates that around 45,000 new jobs could be created by 2010[244][247].

Recyclable waste sorting — Allgemeine Unfallversicherungsanstalt, Austria

As the regulation related to waste was developed primarily for environmental purposes, it does not consider OSH aspects fully[246]. Indeed, in some cases, new waste handling and treating technologies have even increased risks for workers in waste collection, sorting, treatment and disposal activities[248].

The health symptoms observed in workers involved in the management of solid waste are pulmonary, gastrointestinal and skin problems, which have been found to be related to exposure to bioaerosols[152][188][189][244][245][249][250][251][252][253][254][255][256][257][258][259][260]. Management of solid waste includes a multitude of activities, from collection, reception, sorting, recycling of materials, to biological treatment of organic material (for example, composting), thermal treatment (including incineration with energy recovery) and landfill. The handling of medical waste presents extra challenges such as the risk of contamination with sharps[50][248][261][262].

At present, there are very few Occupational Exposure Limits (OELs) for airborne microorganisms or their associated toxins (see '4.2.6 Difficult risk assessment of biological agents in the workplace').

Exposure to biological agents

Workers are exposed to biological agents contained in bioaerosls and organic dust.

Several studies support the thesis that exposure to bioaerosols engenders a multitude of health problems among workers in the collection, processing, recycling and disposal of waste[152][188][189][244][245][249][250][251][252][253][254][255][256][257][258][259][260], especially municipal waste with large amounts of organic material[251][256] and composting-related activities[249].

Workers' exposure to bioaerosols results from the generation of organic dusts[249][263]. For instance, in atmospheres of sorting cabins of recycling plants and of composting plants a nearly constant statistical correlation between the concentration of certain particle fractions of the inhalable dust and the number of airborne moulds in the air has been observed[264]. Biological agents may become airborne as a result of any mechanical manipulation such as transportation or sorting. Landfill workers, for instance, are potentially exposed to high levels of dusts containing microorganisms which can be spread during the dumping or moving of waste[244].

The arising aerosols may include a diversity of airborne microorganisms (bacteria, viruses, mould), the toxic products thereof (endotoxins, mycotoxins, volatile organic compounds (VOCs)) and organic dust[189][245][250][253][254][265][266][267] and generally have a very complex composition, which depends on the type of waste considered as well as the type of waste treatment activity (Table 2).

Table 2. Airborne microorganisms in waste treatment activities[265]

Microorganism	Sorting	Combustion	Composting	Landfill
Bacteria total	+	+	++	+
Streptococcus				+
Enterobacteria				+
Actinomyces	+	+	++	+
Thermoactinomyces	+	+	++	+
Moulds total	++	+	++	++
Aspergillus flavus	+/-	+	+	+/-
Aspergillus fumigatus	+	+	++	++
Aspergillus niger	+	+	++	+
Aspergillus nidulans		+	++	+
Cladosporium spp.	+/-	+	++	+
Penicillium spp.	+	+	++	+

+/-, detectable; +, 103–104 CFU/m^3; ++ >105 CFU/m^3 (CFU, Colony Forming Unit).

The composition of microorganisms and their airborne concentration vary over time.

The composition of microorganisms may vary over time. Indeed, biological agents may influence each other's growth. Moreover, because of their ability to reproduce, small amounts of microorganisms may grow considerably in a very short time under favourable conditions[268]. The composting of organic waste matter, for example, uses the decomposing action of microorganisms to reduce the quantities of biodegradable solid waste sent to landfill and to produce humus or methane. A large number of microorganisms are produced during composting and the number of thermophilic actinomycetes increases considerably during the composting process[269].

Apart from the exposure to the microorganisms themselves, workers may be exposed to their constituents, such as endotoxins released from bacteria[152][153][154][245][254], or to their products, such as microbiological VOCs produced by fungi. Several papers indeed report VOC emission in waste collection[258][270], compost plants[268][271][275][272][273][274][276], landfills and resource recovery plants[257]. These VOCs are both inherent to the waste itself and produced by the microorganisms present in the waste and degrading the organic material. Up to 110 organic compounds from windrow compost have been identified[270][271]. Typical VOCs found in composting plants are carboxylic acids (for example, acetic acid[245]) and their esters, some alcohols, ketones, aldehydes and terpenes[270][274], trichloroethane, toluene, tetrachloroethene and p-xylene[275], d-limonene[245][275], dimethyl sulphide and siloxane[245], as well as further hydrocarbons[245][270]. Among 13 aromatic VOCs found during the composting of the organic fraction of municipal solid wastes, the highest levels were found for toluene, ethylbenzene, 1,4-dichlorobenzene, p-isopropyl and naphthalene[272], they but were still always lower than the OELs[268][270][275][274]. During the first two weeks of storage of biodegradable domestic waste up to 5.0 mg/m^3 methanol, 4.2 mg/m^3 ammonia and 2.8 mg/m^3 hydrogen sulphide were measured[245]. Most VOCs are given off early during the composting process and their production rates decrease over time at thermophilic temperatures[275][272].

Waste from medical treatment of humans or animals, and from biological research also contains biological agents.

Medical or clinical waste is derived from the medical treatment of humans or animals or biological research. Although biological agents themselves are not generally handled in a clinical setting, contaminated waste from patients with infectious diseases is likely to contain biological agents[50][277].

Workers at risk

In the literature reviewed, workers at the following workstations, involved in the following activities or operating the following devices, are identified as being at increased risk of exposure to biological agents in waste treatment activities:

- handling and recycling of solid waste: transfer station, waste unloading, tipping halls or bunkers, front-end loader in tipping hall, conveyor, manual sack opening, manual sorting, visual check, scales, compactors, technical maintenance and servicing of machinery, cleaning[152][188][249][251][255][256][257][258][259][269][278][279][280][281][282], and more specifically

Composting plant — Allgemeine Unfallversicherungsanstalt, Austria

  - in sorting paper activities: paper or cardboard maculation and balers[257][280][281][283];
  - handling of medical waste[50][248][261]: collection, handling and treatment;
- compost plants[249][258][268][271][274][278][279][280][281][284]: loading garbage, mill outlet, control room, siring, pile creation and agitation, shedding, airing and feeding, hand loading of composting, digging waste, turning compost and shaking conveyor in outdoor compost plants, pre- and post-composting loading containers from conveyor and dismantling compost pile in indoor compost plants;
- landfills[257][258]: bulldozer, lorries and cranes, cabins and dumping site;
- incineration processes[282]: all workplaces.

Health outcomes

Human response to exposure to biological agents depends on the specific substance involved, the level of exposure and the individual susceptibility of the exposed worker[254] (see also 'Dose-effect relationship' in '4.2.6 Difficult risk assessment of biological agents in the workplace'). The major occupational health problems encountered in workers involved in waste-treatment plants and composting plants are pulmonary diseases, organic dust toxic syndrome (ODTS), gastrointestinal problems and skin problems, and irritation of the eyes and mucous membranes[153][187][188][189][190][245][250][251][252][259][260][274][281][283][284][285][286][287]. Heavy infections (for example, pneumonias and aspergillosis) are rarely reported and seem to remain individual cases. Those most at risk of developing health complications if exposed to large concentrations of spores include people who already suffer from asthma, immunosuppressed people, or people taking high doses of corticosteroids[250][251][284][285][288][289]. However, because of the increased number of immunosuppressed and therefore susceptible individuals, the presence of airborne Aspergillus fumigatus in the workplace is a concern[288].

Health effects include pulmonary, gastrointestinal and skin problems, and allergic reactions.

Increased risk of gastrointestinal symptoms was found in waste collectors[252], workers handling waste in general and compost workers[281]. A dose-effect relationship was found between nausea and endotoxin exposure, and between diarrhoea and exposure to both endotoxins and viable fungi[254].

Inhalation of biohazards in large quantity can cause transient symptoms, with cough, chest-tightness, dyspnoea, flu-like symptoms such as chills, fever, muscle and joint pains, fatigue and headache. These symptoms are generally termed ODTS and may be caused by toxins of microorganisms[253]. The acute clinical symptoms of ODTS, caused mainly by endotoxins of gram-negative bacteria, occur between six and 12 hours after exposure and last about four hours. In contrast to allergies, which can induce similar clinical symptoms, ODTS is not characterised by the appearance of specific antibodies. ODTS was observed in compost workers with high exposure to bacteria and moulds[285][290]. Over the past few years, it has been observed in different studies that exposure to organic dust (bioaerosols) also leads to obstructive lung diseases, without any allergies being diagnosed[291][292]. Endotoxins[286], mycotoxins[293] or (1–3)-ß-D-glucans[294] are regarded as elicitors.

An increased incidence of upper airway inflammation and respiratory symptoms has been found in waste collectors. Exposure to organic dust probably underlies the inflammation mediated by neutrophils that results in respiratory symptoms[278]. But exposure to a mixture of VOCs with a total of 25 mg/m^3 hydrocarbons can cause irritations in the upper respiratory tract and inflammatory responses in the upper airways[273]. Respiratory diseases are of special occupational interest[288]. Monitoring the lung function has been found to be a useful diagnostic tool to determine long-lasting bioaerosol exposure[295][296]. Indeed, a five-year prospective study found that the lung function of compost workers declined significantly during the observation period as 15% of the workers suffered a reduction of their breathing capacity of more than 10% in five years[297].

Allergic reactions are an increasing problem in industrialised countries. In recycling plants for paper, glass, synthetic and wrapping materials, as well as in composting plants, moulds are risk factors for allergies[248]. There are two types of allergy forms: type I allergy and type III allergy. While type III allergy is especially observed in the waste industry, IgE-mediated type I allergies, for example, asthma bronchiale, were found to be rarer in the population of compost workers than in controls[249][285] — although this may be a consequence of the so-called healthy worker effect. Type III allergy, for example, the extrinsic allergic alveolitis — also called hypersensitivity pneumonitis, mediated by Immunoglobulin G — is caused by long-lasting contact with antigens of fungi and actinomycetes, for example, from Aspergillus fumigatus or Saccharopolyspora rectivirgula[263]. Individual susceptibility also plays an important role in the development of a type III allergy[298].

Inflammation, disturbances of mucous membranes of the eyes and of the respiratory system can be attributed to infectious, allergic or toxic effects. Thereby physical damage of the mucous membranes of the respiratory system by inhalable dust (non-specific particle effect) as well as chemical activation, for example, T-cell activation and invasion of lymphocytes, may lead to toxic pneumonitis or other acute irritation symptoms (mucous membrane irritation (MMI)[189][285][297]). Whether a non-specific particle effect or a specific toxin is the cause could so far not be clarified. A synergism of both is discussed[297].

Last but not least, handling hospital waste and sharps may lead to infections with viruses such as hepatitis and HIV[248].

Possible prevention measures

Prevention should be adapted to the specificities of each waste branch and activities characterised by[246]:

- 'multi-task' workers often involved in several different activities, hence a multi-exposure
- small enterprises often employing low-skill and poorly trained workers
- poor knowledge/complexity of the waste entering into the treatment process
- waste-related technologies and processes in permanent evolution
- waste-treatment activities often installed in long-standing facilities in which it may be difficult to implement collective protection measures.

Although indicative OELs have been established in some Member States for some airborne microorganisms or their associated toxins, there are few obligatory OELs and their establishment is difficult (see '4.2.6 Difficult risk assessment of biological agents in the workplace'). However, technical regulations already exist[299][300][301][302]. While it is not possible to completely eliminate the risks posed by biological agents from waste-related activities, the most efficient prevention measure is to reduce the generation of dust, bioaerosols and VOCs in the workplace[249][297]. Several Member States have already developed preventive measures including the replacement of manual sorting with, for example, mechanical presorting, the installation of sorting cabins with proper ventilation, local exhaust ventilation for sorting lines, closed vehicles equipped with air filters and the use of adequate protective clothing, including proper gloves. Hygiene plans, regular cleaning, and decontamination

There are few OELs and their establishment is difficult. However, technical regulations exist.

measures, have also contributed to a considerable reduction in the exposure of workers[248].

The most efficient prevention measure is to reduce the generation of dust and bioaerosols.

A number of prevention and protection measures — some of which are being tested[303] — are described in the literature[249][254][257][258][264] [268][272][274] [297], the priority being given to collective prevention measures rather than to personal protection measures:

Technical safety measures

- minimisation of the release of bioaerosols, for example:
  - waste processing immediately after delivery; for instance, in bio-compost plants, avoiding long storage times of critical materials by immediate processing reduces emissions[264];
  - enclosing machinery and equipment to reduce bioaerosol immissions;
  - dust cover for unloading waste trucks.
- avoidance of manual processing;
- controlled atmosphere in workplaces with air filtration (high efficiency particulate air (HEPA)) or air conditioning, for example, in closed cabins for manual waste sorting, bulldozer or lorry drivers' cabs, crane cabs; high-level maintenance of ventilation systems[254]; ventilation should be efficient to remove the airborne pollutants[258][257];
- in the composting process, biofiltration in addition to gas treatment units helps to decrease VOC levels and offensive odours commonly found in municipal solid wastes composting facilities[272]; biofiltration was found to reduce the concentration of ammonia, dimethyl disulfide, carbon disulfide, formic acid, acetic acid, and sulphur dioxide (or carbonyl sulphide) by 99, 90, 32, 100, 34 and 100%, respectively[304].

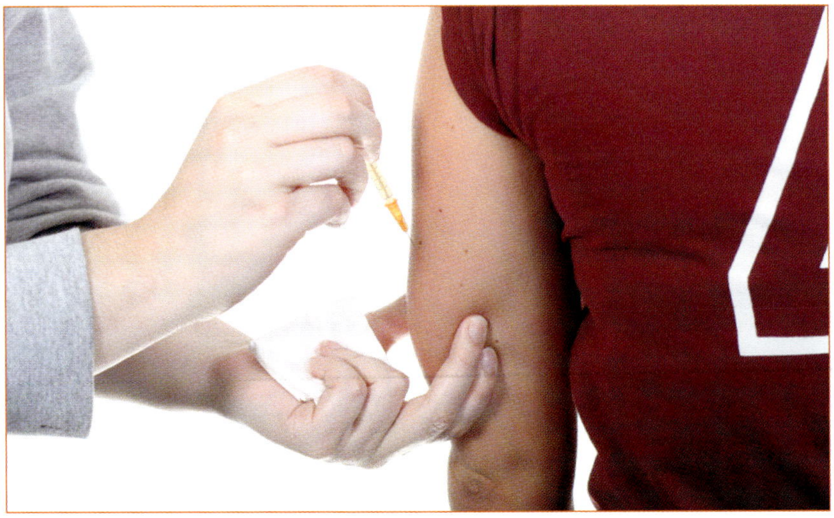

Vaccine injection

Organisational safety measures

- isolation (for example, with automatically closing doors, sluice) of workstations with bioaerosol emissions from all other working areas
- restriction of entrance to areas with high bioaerosol levels to an operational minimum number of workers[274][305];
- workers' training, which plays a crucial role in promoting safe working habits[258][257];
- preventive medical check-ups and vaccinations (tetanus).

Hygienic measures[268][257]

- cleaning of workplaces must be considered an integral part of operations and it should be carried out properly in order to minimise dust generation; workplaces should be designed with easy-to-clean surfaces;
- separate storage of private and working clothing;
- scheduled regular cleaning and changing of working and protective clothes;
- provision of facilities to wash hands when leaving the workplace, especially before entering staff room, and to shower after shift;
- avoidance of contact of eye, nose and mouth with unwashed hands;
- avoidance of eating, drinking or smoking at the workplace and provision of clean and separated storage facilities for food and drinks.

Personal safety measures[258][274][306]

- respiratory protection against bioaerols and VOCs including respirators with clean filtered air supply (in case of asthmatic workers or very high exposure);
- protective clothes, gloves and goggles.

Specifically for medical and clinical waste[50][261][262]

- policy which ensures safe collection, storage, transport and final disposal of such waste, including the safe management of sharps;
- incinerate or autoclave infectious waste before disposal.

4.2.6. Difficult risk assessment of biological agents in the workplace

Food industry — Allgemeine Unfallversicherungsanstalt, Austria.

As biological agents are rarely visible, the risks they pose to workers are not always considered adequately.

Biological agents are defined in Directive 2000/54/EC on the protection of workers from risks related to exposure to biological agents at work[307] as microorganisms — i.e. bacteria, viruses and fungi (yeasts and moulds), including those which have been genetically modified — cell cultures and human endoparasites, which may be able to provoke any infection, allergy or toxicity. They are found in many sectors and workplaces. However, as they are rarely visible, the risks they pose to workers are not always considered adequately. Exposure to biological agents in the workplace may result directly from the work — this is for instance the case of a microorganism culture in a microbiology laboratory or of the use of microorganisms in the food industry — or may be incidental to it, as in farming or waste treatment activities[50][248].

Directive 2000/54/EC lays the principle for the assessment and management of biological risks in the workplace.

It is within the scope of Directive 2000/54/EC to determine and assess the risks posed by biological agents in the workplace. This directive is to be applied to any activity where workers are actually or potentially exposed to biological agents as a result of their work. According to this Directive and Directive 93/67/EEC[308], the risk assessment shall entail:

- a hazard identification, which consists in identifying the biological agents present and the adverse effects that they have an inherent capacity to cause;
- a dose (concentration) — response (effect) assessment, which is the estimation of the relationship between the level of exposure to a substance and the incidence and severity of an effect;
- an exposure assessment, which is the determination of the concentrations, routes of exposure, potential for absorption, and the frequency and duration of exposure, in order to estimate the doses to which workers are or may be exposed;
- and a risk characterisation, which is the estimation of the incidence and severity of the adverse effects likely to occur in workers due to the actual or predicted exposure to a substance.

If workers are exposed to several groups of biological agents, then the risk must be assessed in terms of the dangers posed by all the hazardous biological agents present. This risk assessment must be renewed regularly and when working conditions change in a way that affects the workers' exposure to such biological agents[309].

Guidance and standards are available to help in the risk assessment of biological agents[310][311][312][313][314][315]. However, the assessment of biological risks is seriously hampered as, contrary to the majority of chemical and physical factors, neither commonly approved criteria for assessing exposure to biological factors, nor well-established dose-effect relationship and occupational exposure limits (OELs) are — at the time of writing of this report — yet available[191][193][194][316][317]. When a work activity involves the intentional use of biological agents, such as in a microbiological laboratory, the biological agent will be known and can be monitored more easily. However, when the occurrence of the biological agents is an unintentional consequence of the work — for instance in waste sorting or agricultural activities — the assessment of the risks to workers is more difficult[248].

Measurement methods and exposure assessment

The assessment of exposures to biohazards offers distinct challenges from those for inorganic aerosols and chemical agents. The difficulties are mainly linked to the fact that measurement methods do not yet enable the efficient detection and measurement of biological agents. A number of methods for sampling and analysing biological agents and estimating occupational exposures, as well as recommendations on suitable measurement strategies, are described and evaluated in scientific papers.

Airborne exposure to biological agents in the environment can be assessed by counting culturable microorganisms in air samples or in settled dust samples. The culture medium is formulated to test for broad-spectrum bacteria and fungi or to select for specific groups, genera or species. These methods can be based on microscopic, microbiological, biochemical, immunochemical or molecular biological analysis. Various culture methods used to determine the bacteria concentration in air samples are reviewed in the literature[318]. Culture-based techniques are important and often-used tools for the risk assessment of biohazards[319][320], although they provide rather qualitative analysis. Moreover, culture methods have been proven to be of limited use in population-based studies[193][317][321].

Non-culture methods and assessment methods appear more reliable. However, experience with these methods is still generally limited[193]. Sampling of non-culturable bioaerosols can be based on air filtration or liquid impinger methods[322][323][324][325][326]. Fast and sensitive techniques based on light scattering spectrometry can be used to determine microorganism presence in liquid samples[327]. Electron microscopy or scanning electron microscopy could also be used and allow better determination of biological agents. Simple light microscopy may be used to count microorganisms. However, counting is based only on morphological recognition, which may result in severe measurement errors, as dead microorganisms are not differentiated from living ones[323][328][329][330].

There is a need to develop reliable tools and methods for the measurement of airborne biological agents.

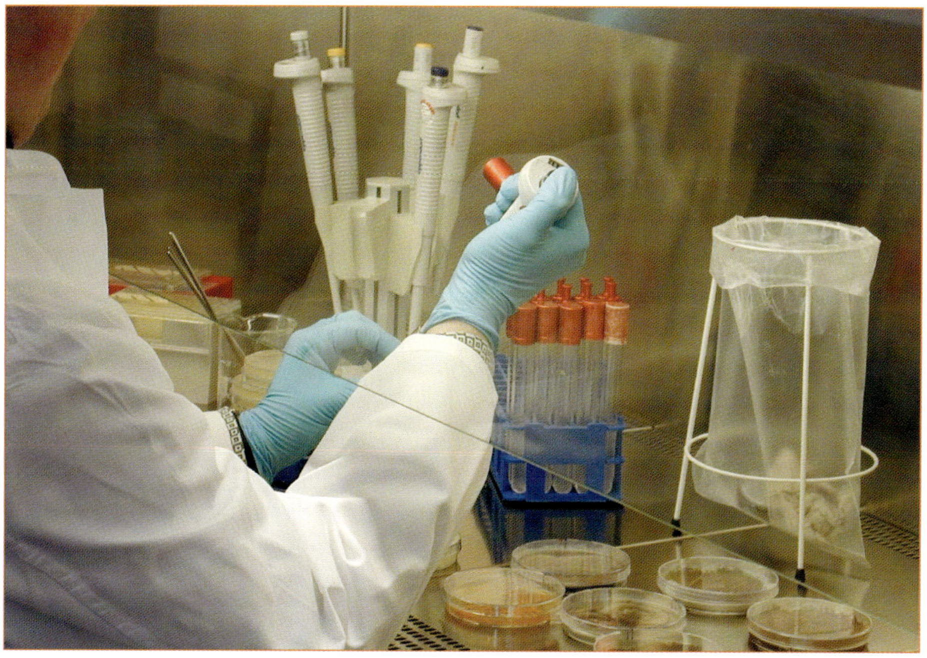

Characterisation of microorganisms in samples — Berufsgenossenchaftliches Institut für Arbeitsschutz, Germany.

Instead of counting culturable or non-culturable microorganisms themselves, their constituents or metabolites can be measured as indicators of biological exposure[193]. Some cellular components may be used as markers of groups of microorganisms. VOCs produced by fungi may be suitable markers of fungal growth[331]. The assessment of markers such as polysaccharides and ergosterol based on gas chromatography-mass spectrometry or specific enzyme immunoassays could also allow identification of fungal biomass[332][333]. Important methods based on quantitative polymerase chain reaction (PCR) are emerging for the identification and quantitative assessment of specific airborne biological agents[191][334][335]. Some biological agents such as endotoxins, besides being measured because of their own toxic potency, may also be markers of other biological agents. Endotoxins can be measured using a Limulus amoebocyte lysate (LAL) test or methods employing gas chromatography-mass spectrometry[191][336][337][338]. Some methods of measuring ß(1?3)-glucans are based on the LAL assay and on an enzyme immunoassay[339][340]. However, the quantitative relation between biomarkers and airborne exposure has not been sufficiently recognised yet. Only a few biomarkers, mainly endotoxins, have been identified and validated[326]. No direct methods to measure biological agents or metabolites in body fluids have been described. Therefore, results of measurements of biological agent concentrations should be interpreted with caution[341][342].

Antibody-based immunoassays, particularly enzyme-linked immunosorbent assays (ELISA) are widely used for the measurement of bioallergens in settled dust in buildings[343][344][345][346]. Methods for assessing exposure to bioallergens from different sources have also been published[347][348][349][350][351]. As for viruses, exposure assessment has hardly been developed for occupational environments[193].

Exposure assessment has been helped by advances in assays for microbial agents — such as the measurement of endotoxin or glucans with a LAL assay, of allergens with an enzyme immunoassay, or of fungal extracellular polysaccharides, which are more established methods thanks to the improved stability of most of the measured components, allowing longer sampling times for airborne measurements, and the better possibility to test for reproducibility[191][352][353][354]. However, even these methods still show significant variations in exposure assessment between laboratories and are only poorly validated, and often not even commercially available[193].

In fact, there are a number of difficulties with the measurement methods described above. They generally imply complicated procedures and calculations, as well as incurring high costs per sample[317][322][326][355]. Some difficulties are also related to taking representative samples. Indeed, biological agents may influence each other's growth. Moreover, because of their ability to reproduce, small amounts of microorganisms may grow considerably in a very short time under favourable conditions[248]. For this reason, and also because of seasonal variations in airborne levels of microorganisms, repeated sampling is often required to allow for an accurate determination of airborne microorganism concentrations, but it is expensive and therefore difficult to implement widely. Furthermore, some microorganisms and spores are extremely resilient while others may be easily degraded in the sampling process. The issues of storage and transport of bioaerosol samples are often not addressed, although these conditions may affect the activity of some biological agents. Additionally, high concentrations of bioaerosols in many workplaces require the integration of multiple samples[193][248]. In the face of such complexity, meaningful quantitative sampling methods are frequently unavailable. Additionally, these methods have proven unable to detect all possibly relevant biological agents present in the workplace[317][322][326][355]. This means that the complete range of airborne organisms may not be recognized and the true concentration of biological substances may be miscalculated[191][321][356][357][358].

In conclusion, the measurement methods available, even the more established ones, have not yet been fully validated and routinely applied[317][322][326][352][353][354][355]. There is a need to develop adequate sampling techniques and analytical methods further to allow for better identification and for quantitative exposure assessment of biological agents in the workplace, which are essential steps towards a proper risk assessment.

Dose-effect relationship

Dose-response relationships have not been established for most biological agents. One obvious reason for this is the lack of valid quantitative exposure assessment methods highlighted previously[193][254].

Moreover, the precise role of biological agents in the development or aggravation of symptoms and diseases is only poorly understood. Human response to exposure to biological agents depends on the specific material involved and individual susceptibility[254]. In most situations, combined exposure to complex mixtures of toxins and allergens, as well as interactions with non-biological agents, occur in the workplace and a wide range of potential health effects have to be considered.

However, it is difficult to determine which of the constituents primarily accounts for presumed health effects[193].

Human response to biological agents depends on the material involved and individual susceptibility.

Furthermore, there is very little information on the 'infection doses' or 'relevant concentrations' of the biological agents that inevitably cause diseases. Some pathogenic microorganisms may be hazardous at extremely low levels while other organisms may only become important health hazards at orders of magnitude of higher concentrations. Additionally, the susceptibility to exposure varies with the individual's genetic predisposition, age, state of health, and with concurrent exposures[193][309].

Indeed, several studies have revealed a complex, dose-dependent, non-linear relationship between environmental exposure to some biological agents — such as endotoxins, fungal spores and other pathogen-associated molecular patterns (PAMPs) — and the outcome of immune responses. Exposure to such agents may play a critical role in the pathogenesis of complex diseases — such as asthma, atopy, respiratory allergies and sensitisation to allergens — and result in different responses in humans depending on the environmental context and on the interplay between environmental exposure and genetic background of individuals. In fact, they have been found to induce, but conversely to also protect from, these diseases[128][174][175][176][177][178][179].

Important areas that require further research for the establishment of dose-response relationships with regard to biological agents therefore include the inter-individual susceptibility for biological exposures, the potential protective effects of microbial exposures, as well as the interactions of bioaerosols with non-biological agents[193].

Recommendation for Occupational Exposure Limit (OEL) values

The large uncertainties in quantitative concentration assessments and the lack of established dose-effect relationships hamper the development of legal OELs[193][359][360]. The legislative framework sets OELs for biological agents for only some toxins in some Member States[248], or for contaminants, such as wood dust, subtilisins (bacterial enzymes) and flour dust. In the Netherlands, for instance, there is a legal OEL for aflatoxins set at 0,005 m/m^3[361]. Exposure limits for bacterial endotoxins have been proposed but not yet definitively established (see '4.2.3 Occupational exposure to endotoxins'). A number of papers reviewing proposals for OEL settings and formulating recommendations are available[193][195][196][254][359][360].

The establishment of OELs is hindered by the unreliable methods for measurement of biological agent concentrations and, conversely, the lack of OELs renders the risk assessment of biological agents difficult. Indeed, OELs offer guidance on how to interpret the results of exposure assessment with a view to evaluating the severity of the risk[317]. Some recommendations for decision-making and interpretation of measurement results, without OELs being available, are proposed[359][360].

Conclusion

It is within the scope of Directive 2000/54/EC to determine and assess the risks that are posed by biological agents in the workplace. This directive should therefore be applied to any activity where workers are actually or potentially exposed to biological agents as a result of their work. However, knowledge of biological hazards is still relatively scarce. In order to enable a proper exposure assessment, there remains a clear need for research to develop better tools for the detection and measurement of biological agents — particularly based on non-culture techniques since culture methods have been proven to be of limited use. A crucial area for further development is the validation of available measurement methods and their international harmonisation to reduce interlaboratory variability. This would enable us to define commonly approved criteria and accepted protocols for assessing exposure to biological hazardous substances — including concise and uniform guidelines on sampling, storage, extraction and analytical procedures — and to better understand the relationships between exposure and occupational health effects. This would facilitate the establishment of OELs which, conversely, would support the proper interpretation of measurement results in a risk assessment procedure.

5.

COMPLETE RESULTS OF THE SURVEY

In the following sections, the exact descriptions of the risks rated by the experts are listed in tables together with the number of respondents, the mean value of the ratings and the standard deviation. These figures are also compiled in diagrams. When available, the comments added by the respondents to the items are listed to provide some context and support to the ratings.

5.1. Substance-specific biological risks

Diagram 6. Substance-specific biological risks identified in the survey

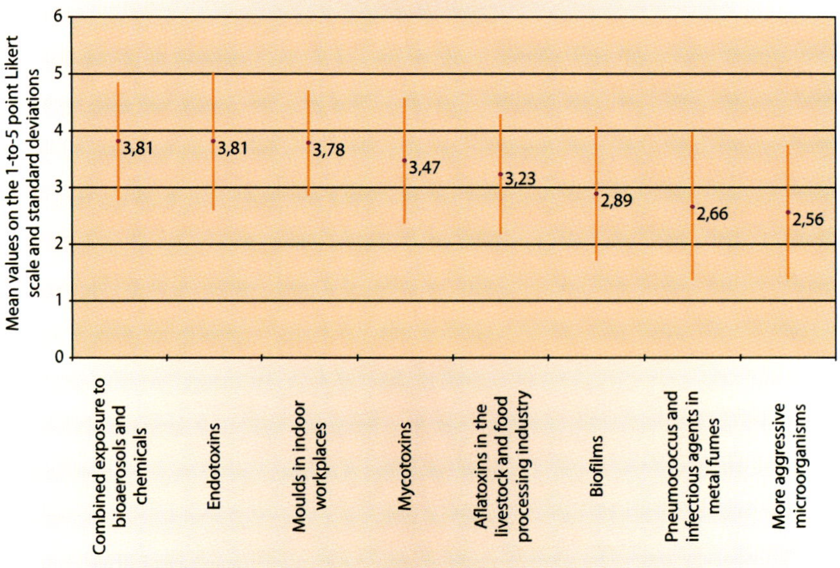

Half of the items (four of eight) mentioned in the survey referring to specific biological agents are agreed as emerging risks by the experts (MV>3.25): 'Combined exposure to bioaerosols and chemicals', as well as occupational exposure to 'endotoxins' (see literature review '4.2.3 Occupational exposure to endotoxins'), to 'moulds in indoor workplaces' (see literature review '4.2.4 Moulds in indoor workplaces') and to 'mycotoxins'.

Two items that had been proposed by some experts in the first brainstorming round of the survey have been eliminated in the subsequent rounds and rated as non-emerging risks (MV<2.75): 'Pneumococcus and infectious agents in metal fumes', and occupational exposure to 'more aggressive microorganisms in laboratories'.

It should be noted that the consensus on the ratings among the respondents is especially low for the items 'endotoxins' and 'pneumococcus and infectious agents in metal fumes' (SD>1.2).

Table 3. Prioritised list of substance-specific biological risks identified in the survey (N=number of experts answering the specific item; mean value; standard deviation)

MV>4: risk strongly agreed as emerging
3.25<MV≤4: risk agreed as emerging
2.75≤MV≤3.25: status undecided
2≤MV<2.75: risk agreed as non-emerging

NB: None of the risks was strongly agreed as non-emerging (MV<2).

Substance-specific biological risks	N	Mean Value (MV)	Standard Deviation (SD)
Bioaerosols and chemicals, the combined effects of which have been very little studied but lead to allergies. More knowledge will help identify the real multi-factorial causes of symptoms for which mono-causal explanations have been made so far.	36	3,81	1,037
Endotoxins: High concentrations in various industrial settings (e.g. in workplaces exposed to organic materials (straw, wood, cotton dust), waste treatment, poultry houses, swine confinment buildings) leading to asthma, loss of lung function, etc.	36	3,81	1,215
Moulds in indoor workplaces due to new construction methods and materials, to the aim of saving energy, and to the lack of maintenance: Exposure to fungal spores for office workers and especially workers involved in building restauration, leading to sensitization and allergies.	36	3,78	0,929
Mycotoxins: Increasing risk as mycotoxins have increasing possibilities to grow in occupational settings, for example, in waste treatment jobs due to the increase of garbage quantities. Potential health effects: cancers, immune deprivations and congenital abnormalities. Groups more at risk: workers in waste treatment occupations, textile and food-processing sectors, and workers involved in wet work.	36	3,47	1,108
Aflatoxin exposure of staff in food processing plants and animal feeding plants may lead to,e.g. cancer.	35	3,23	1.060
Impact of biofilms on healthe.g. in water and air systems.	35	2,89	1,183
Pneumococcus and various infectious agents from metal fumes, the effects of which were previously unrecognised.	35	2,66	1,305
Exposure to potentially more aggressive microorganisms and products in the workplace, mainly resulting from the increasing enzyme use.	34	2,56	1,186

Experts' comments

Risks agreed as emerging (3.25 < MV ≤ 4)

- Combined exposure to bioaerosols and chemicals

There is still a lack of knowledge on the health outcomes of combined exposure to bioaerosols and chemicals.

- Endotoxins

High concentrations of endotoxins can be found in the agricultural sector (swine breeding and grain harvest). One issue often ignored is that the use of bactericides to eliminate bacteria from a contaminated area may actually result in the emergence of other endotoxin-producing organisms in the very same area, maybe resulting in the presence of endotoxins resistant to the bactericide used, and in concentrations even higher than the initial level. However, endotoxins are not always harmful to human health; on the contrary, in some cases they have proven to have positive effects on health. Indeed, exposure to high concentrations of endotoxins may reduce the incidence of allergic reactions such as atopic asthma and allergies through an effect on the balance of T-helper cells 1 and T-helper cells 2. However, more research is needed on occupational exposure to endotoxins and on the dose-effect relationship.

- Moulds in indoor workplaces

Due to financial considerations, builders often do not allow building materials long enough to dry out, which may result in mould growth in the finished building. Indoor mould growth is not only a problem with new buildings, however, as it is also found in older structures. According to one respondent, moulds are only moderate allergens, and only high airborne concentrations of mould spores may cause allergies. A further expert adds that mould-related allergies in workers seem to be an issue for construction workers rather than for office workers. However, another expert mentions that an increase in the number of mould-related occupational diseases is seen in some countries' statistics and adds that, in Finland in 2002, there were 264 cases of occupational diseases caused by moulds, mostly allergies (155 cases). The most common field for these occupational diseases was health care and the social sector with 71 cases, followed by public administration (49 cases), agriculture (43 cases) and education (42 cases). The construction sector reported only seven cases of occupational diseases caused by moulds. This divergence of opinions may be due to geographical differences, to the different level of awareness of mould-related health problems in the different countries, and to the differences in national recognition systems for occupational diseases. Last but not least, it is suggested that exposure to mouldy working environments due to water-damaged buildings may cause health symptoms that are sometimes mis-diagnosed as flu-like diseases.

- Mycotoxins

Occupational exposure to mycotoxins, including aflatoxins, is considered to be an increasing risk because the number of workplaces where workers may be exposed, such as in the waste treatment sector, is increasing. Workers in the agricultural sector are also at increased risk; for instance, when working in grain silos. However, whether exposure to mycotoxins has only harmful effects on human health is controversial as there is evidence that the normal maturation of the immune system requires the stimulation of transmembrane proteins, termed 'toll-like receptors' (TLRs), which are activated by exposure to commonly occurring microbial elements, including endotoxin, mycobacterial lipopeptides and fungal glucans. More research has to be carried out on mycotoxins — more particularly on occupational exposure to airborne mycotoxins — among others to allow for a better understanding of the dose-response relationship.

Undecided (2.75 ≤ MV ≤ 3.25)

- Aflatoxins in livestock and food processing industry

Aflatoxins belong to group 1 carcinogens as defined in the International Agency for Research on Cancer (IARC) classification. There is still a need for more research into aflatoxins and occupational exposure, in particular with a view to supporting the risk assessment of aflatoxin-related risks in the workplace.

- Biofilms

The growth of biofilms results mainly from the poor maintenance and poor cleanliness of air and water systems.

Risks agreed as non-emerging ($2 \leq MV < 2.75$)

- Pneumococcus and various infectious agents in metal fumes

It seems that this item was not correctly formulated in the first survey round, as it is unlikely that microorganisms and infectious agents survive in metal fumes. Maybe the original proposal meant pneumococcus and various infectious agents in metal *fluids* instead of metal *fumes*. However, regarding exposure to metal *fumes*, there is evidence that inhalation of metal fumes increases workers' susceptibility to pneumonia.

- More aggressive microorganisms

According to the experts, microorganisms are not necessarily more aggressive but many of them are found to be opportunistic pathogens (i.e. they cause a disease in a compromised host that typically would not occur in a healthy host), while the number of immunocompromised individuals is growing.

5.2. Workplace and work-process specific biological risks

Diagram 7. Workplace and work-process specific biological risks identified in the survey

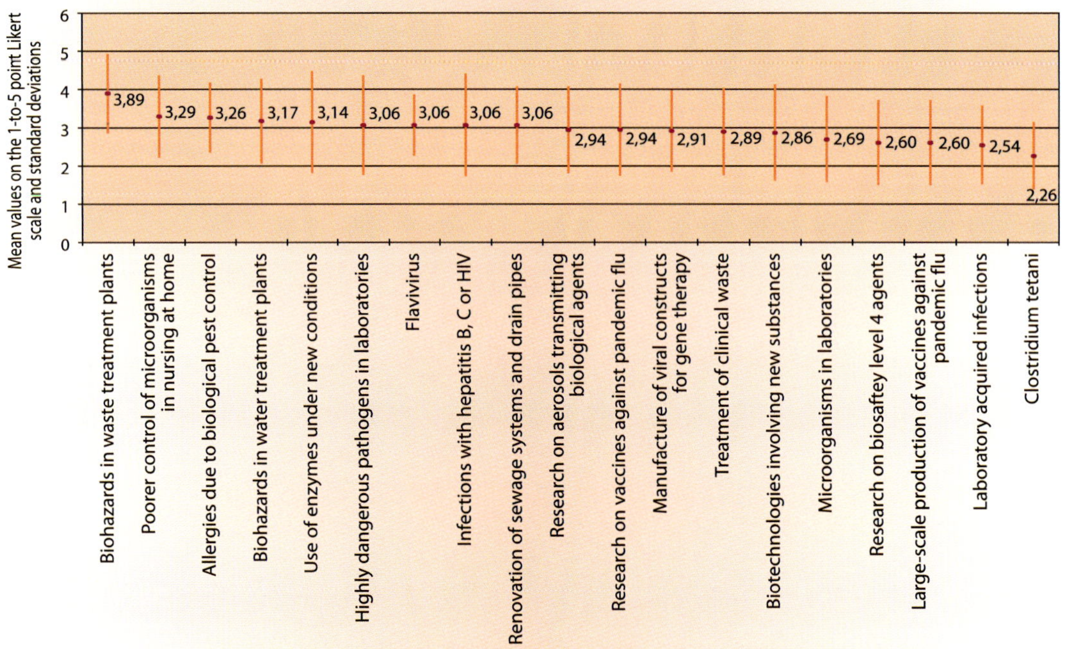

Only three out of the 18 risks specific to certain workplaces and work processes brought up in the first survey round have been rated as emerging.

The occupational risks linked to exposure to 'biohazards in waste treatment plants' are perceived as emerging (3.25<MV≤ 4) and belong to the 10 main emerging risks identified in the survey (Diagram 5). (See literature review '4.2.5 Biological risks in the management of solid waste'). In a complementary expert survey on emerging chemical risks[43], occupational risks related to waste treatment activities have also been rated as strongly emerging (MV=4.11). The two very similar mean values in both surveys may be considered to validate the forecast.

Two further risks are agreed as emerging: one in the health care sector, as a consequence of the poorly controlled conditions medical staff are exposed to when working in patients' homes, especially with a view to exposure to biological agents; and one in the agricultural sector linked to the use of biological pest control.

While the respondents remain undecided (2.75≤ MV≤ 3.25) as to the status of 11 items which had been proposed as potential emerging risks in the first survey round, five items have been clearly rated as non-emerging — three of them concerning risks posed by biological agents in laboratory and research work.

[43] European Agency for Safety and Health at Work, 'Expert forecast on emerging chemical risks related to occupational safety and health'. The report will be published in 2007.

It should be noted that there is particularly little consensus among the experts (SD>1.2) on the items 'infections with hepatitis B, C or HIV in health care staff, police officers and prison staff', 'use of enzymes under new conditions', 'highly dangerous pathogens in laboratories', 'biotechnologies' and 'research on vaccines against pandemic flu'.

Table 4. Prioritised list of biological risks specific to certain workplaces and work processes (N=number of experts answering the specific item; mean value (MV); standard deviation (SD))

MV>4: risk strongly agreed as emerging

3.25<MV≤4: risk agreed as emerging

2.75≤MV≤3.25: status undecided

2≤MV<2.75: risk agreed as non-emerging

NB: None of the risks was strongly agreed as non-emerging (MV<2).

Emerging risks due to certain workplaces and work processes	N	Mean Value (MV)	Standard Deviation (SD)
Biohazards in waste treatment plants (e.g. selective sorting, manufacture of compost) leading to allergies, infectious diseases (bacteria, viruses), toxinic diseases (endotoxins, mycotoxins) and cancers (oncogens). Especially in composting facilities, where there are a wide variety of microorganisms present at the different stages of the composting process, the risks are not completely identified yet.	36	3,89	1,036
Increase of nursing at home — because of pressure on medical budgets — leading to exposure of (less well trained self-employed) medical staff to infectious microorganisms, all the more because the environmental working conditions are not controlled as well as in hospitals.	35	3,29	1,073
Biological pest control in greenhouses leading to allergies.	35	3,26	0,919
Increased number of water treatment plants implying a larger number of workers exposed to risks of allergies, infectious diseases (due to bacteria, viruses), toxinic diseases (due to endotoxins, mycotoxins) and cancers (due to oncogens).	36	3,17	1,108
Use of enzymes under new conditions in the food and detergent sectors (wider and more concentrated applications) leading to respiratory and dermal allergies.	35	3,14	1,332
Exposure to flavivirus in forestry occupations leading to encephalitis.	35	3,06	0,802
Increased need in renovation of old degrading sewage systems and drainpipes in Europe, which are sources of many infectious agents (hepatitis A, endotoxin from gram negative bacteria).	36	3,06	1,013
Increasing number of laboratories handling highly dangerous pathogens (because of potential biothreats, epidemic strains) while not always up-to-date from a biosafety point of view (relatively low level of control by authorities especially in academic settings).	35	3,06	1,305
Increase in infections with hepatitis B, C, HIV in health care sector, police and prison staff as well as in other workplaces.	35	3,06	1,349
Research work on the pathogenicity of aerosols transmitting biological agents, such as those responsible for tubercolosis and SARS, leading to increasing transmission to research workers.	35	2,94	1,136
Vaccines against pandemic flu: Potential for evolutionary drift to produce a novel strain with an antigenic profile for which there is no background immunity (e.g. reassortment between a circulating flu virus and a H5 antigen).	34	2,94	1,205

Emerging risks due to certain workplaces and work processes	N	Mean Value (MV)	Standard Deviation (SD)
Manufacture and applications of viral constructs for gene therapy, which concern more staff in the manufacturing and health care sectors, with the following safety issues: safety of the rDNA construct, safety of the viral vector, pathogenicity, recombination events with viral sequences in the host, safety of packaging cell line, and regulation of gene expression of the rDNA product.	35	2,91	1,067
Handling and processing of clinical waste.	36	2,89	1,141
Biotechnologies, involving new substances in occupational settings (e.g. food production).	35	2,86	1,264
Exposure to different microorganisms in laboratory workplaces.	35	2,69	1,132
Increasing amount of research work on biosafety level 4 (BSL4) agents in order to determine their pathogenicity, likely to lead to contamination of laboratory workers.	35	2,60	1,117
Large-scale production of vaccines against pandemic flu: Conflict in requirements between positive pressure from safety regulations (to work under controlled safety conditions) and negative pressure from public health protection (to produce the vaccines as fast as possible).	35	2,60	1,117
Laboratory acquired infections, regardless of biosafety level.	35	2,54	1,039
Exposure to clostridium tetani, potentially leading to death, in the agriculture sector or leather and fur processing occupations.	35	2,26	0,886

Experts' comments

Risks agreed as emerging (3.25 < MV ≤ 4)

- Poor control of biological risks in nursing at home

Home-nursing is an expanding service activity — to some extent as a result of the ageing population requiring more medical care — performed by a growing number of often poorly trained, self-employed nursing staff, facing increased risks of exposure to biological agents as OSH conditions are more difficult to control in homes than in hospital settings. One respondent indicates that medical staff working under time pressure — which is generally the case with home-nursing staff — are at increased risk of needle stick injuries. Another expert, however, highlights the positive aspects of home-nursing that contribute to reducing the occurrence of nosocomial infections.

- Allergies to biological pest control

The experts comment that the risks posed by biological pest control are of a chemical, rather than a biological, nature.

Undecided (2.75 ≤ MV ≤ 3.25)

- Biohazards in water treatment plants

In the Netherlands, waste water has been treated for approximately 20 years and protection measures for workers have been developed. However, the increasing number of water treatment plants means a growing part of the workforce is being exposed to the biological agents present in the treatment process. In addition to the increasing number of workers, exposure to biological agents in waste water treatment plants is seen as an emerging risk because in more and more cases treatment takes place in confined spaces — as a consequence of water treatment plant design — which results in workers being exposed to higher concentrations of hazardous substances.

Bakery — Central Labour Inspectorate, Ministry of Economics and Labour, Austria

- Use of enzymes under new conditions

The use of enzymes is an issue mainly in the food sector — for example, in meat-processing activities or in baking activities in small- and medium-sized enterprises (SMEs) — as well as in jobs producing or using detergent/laundry products. One respondent points out that, according to Directive 2000/54/EC([44]), enzymes are not seen as biological agents but as chemical substances. Another expert commented that workers in the industrial production of enzymes might be exposed to the biological agents used in the process.

- Flavivirus

Flavivirus is found only in certain geographic areas. One expert also mentions the risk of infection with Hanta virus (mainly transmitted by rodents)([45]). However, another expert points to the fact that, according to Directive 2000/54/EC, the hepatitis C virus, which is ubiquitous, is also classified as a flavivirus.

- Highly dangerous pathogens in laboratories

Although the handling of highly dangerous pathogens in laboratories is better controlled in European countries than outside, more resources need to be made available for a better surveillance of the related risks.

- Infections with hepatitis B, C, HIV

Some prevention measures aimed at reducing the number of infections contracted at work in the medical sector, in the police and in prisons have already been developed. However, the number of cases of work-related tuberculosis, for instance, is still increasing for these occupations.

- Renovation of old, degrading sewage systems and drainpipes

New techniques for the renovation of old sewage systems and drainpipes in Europe are available, and protection measures have been developed for workers carrying out these activities. However, biofilms that develop within the sewage systems and drainpipes may still pose important health risks to workers involved in these activities.

- Research on aerosols transmitting biological agents, such as those responsible for tuberculosis and SARS

Policies for the containment of biohazards, as well as good laboratory practices (GLPs) have been developed. In addition, researchers involved in this type of work should be trained on the risks. Furthermore, all research equipment should be adequately maintained, and its reliability strictly controlled, in order to minimise the risk of worker

([44]) 'Directive 2000/54/EC of the European Parliament and of the Council of the 18th September 2000 on the protection of workers from risks related to exposure to biological agents at work (seventh individual directive within the meaning of Article 16(1) of Directive 89/391/EEC)', *Official Journal* L 262, 17th October 2000, pp. 21-45, http://eur-lex.europa.eu/LexUriServ/LexUriServ.do?uri=CELEX:32000L0054:EN:HTML

([45]) 'Hantaviruses (Puumala, Hantaan, Sin Nombre and others) infecting field rodents may be a cause of hemorrhagic fever with renal syndrome (HFRS) or pulmonary syndrome (HPS) in farmers and laboratory workers.' Taken from: Dutkiewicz J., 'Occupational bio hazards: current issues', *Medycyna pracy*, Vol. 55, No 1, pp. 31-40, 2004 contamination. Additionally, staff other than researchers working in bio-research facilities (such as maintenance workers) are also at risk and should be included in risk assessment and prevention activities.

- Viral constructs for gene therapy

Workers manufacturing viral constructs for gene therapy, as well as medical staff using these, are at risk. There is a need for a thorough risk assessment and for workers' training.

Risks agreed as non-emerging (2 ≤ MV < 2.75)

- Research on biosafety level 4 (BSL4) agents

According to some experts, policies on biohazard confinement, good laboratory practices (GLPs), and the mandatory workers' training are tools that are already available and help to control the risks related to research activities on BSL4 agents. An expert also notes that there is a regulatory permission system in place to regulate such activities. One expert mentions that research on BSL 2 and 3 agents actually puts workers more at risk because they are less controlled activities with a consequent lower level of workers' awareness for the related risks.

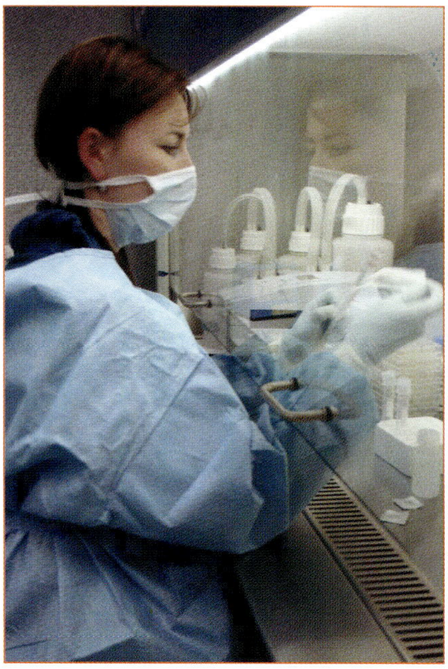

Strict procedures are used to manipulate viruses in BSL 3 laboratories – European Commission

- Large-scale production of vaccines against pandemic flu

Effective technical solutions exist to help prevent the risks posed to workers by the large-scale production of vaccines against pandemic flu. However, it is often a question of the financial resources available whether these solutions are implemented or not. What is more, in some cases, there may be a conflict between OSH and public health requirements to produce the vaccines under better-controlled, safer working conditions on the one hand, but as fast as possible on the other.

- Laboratory-acquired infections

There is still a need for research on laboratory-acquired infections, as well as for more efforts to be made to provide risk assessments at the workplace and appropriate workers' training. At present, laboratory-acquired infections are best controlled for activities involving BSL4 agents.

- Clostridium tetani

Clostridium tetani, which may lead to death, is mainly found in the agriculture sector and in leather and fur processing jobs. A cheap and effective vaccine without any contra-indication has already been available for many years. Making this vaccination compulsory for workers at risk would be a good initiative. In the case of occupational injuries causing dirty wounds (such as injury with rusty nails in renovation works), prompt use of antiseptics and the early start of antibiotics are advocated. This is also true for other microbes that may be present in dirty wounds. One respondent points to the fact that, beside clostridium tetani, workers may also be exposed to anthrax in the agriculture sector and leather and fur processing occupations.

5.3. Biological risks resulting from poor risk management and prevention practices

Diagram 8. Biological risks resulting from poor risk management and prevention practices identified in the survey

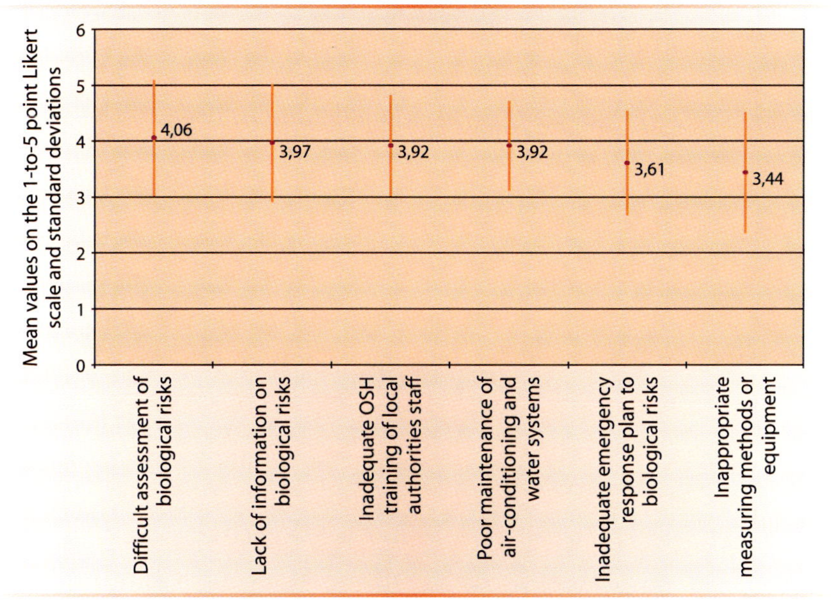

Risks engendered by poor risk management seem to play a major role, since all six items brought up in the first brainstorming survey round have been rated as emerging risks in the later survey rounds.

Whilst, according to Directive 2000/54/EC, employers have the duty to determine and assess the risks posed by biological agents in any workplace where workers are actually or potentially exposed to these, in practice the proper assessment of biological risks still remains difficult (see literature review '4.2.6 Difficult risk assessment of biological agents in the workplace') and is strongly agreed (MV>4) to be an emerging risk in itself by the respondents to the survey. The 'lack of information on biological risks in some workplaces' and the 'inappropriate methods' for the measurement of biological agents, which make risk assessment difficult, are identified as emerging risks (3.25<MV=4). The consistency in the respondents' evaluation of several items linked with difficulties in assessing biological risks may be considered to validate the forecast. It is also interesting to note that exposure assessment of biological agents has been identified as a priority in a review of various national, European and international resources identifying future research needs in the field of OSH([46]).

As well as deficient risk management, inadequate or lacking preventive measures, such as the inadequate provision of OSH training to workers — especially in local

([46]) European Agency for Safety and Health at Work, 'Priorities for occupational safety and health research in the EU-25', Luxembourg, 2005 authorities — poor maintenance of equipment, water and air systems, or inappropriate or missing emergency preparedness and plans, are also considered to pose emerging risks to workers.

Table 5. Prioritised list of biological risks resulting from risk management and poor prevention practices identified in the survey (N=number of experts answering the specific item; mean value (MV); standard deviation (SD))

MV>4: risk strongly agreed as emerging
3.25<MV≤4: risk agreed as emerging
2.75≤MV≤3.25: status undecided
2≤MV<2.75: risk agreed as non-emerging

NB: None of the risks were strongly agreed as non-emerging (MV<2).

Biological risks related to risk management and prevention practices	N	Mean Value (MV)	Standard Deviation (SD)
Poor or difficult assessment of biological risks.	36	4,06	1.040
Lack of information on biological risks in different workplaces (e.g. in offices workplaces oragriculture).	36	3,97	1,055
Inadequate training, poor knowledge of OSH or even poor basic awareness of risks of local authorities staff (e.g. sewage, excavations or waste collection jobs, etc.).	36	3,92	0,906
Poor maintenance of air-conditioning (whose use is increasing) and of water systems (e.g. legionella, aspergilosis in hospitals). New knowledge about the presence of legionella will help the correct diagnose of symptoms so far wrongly attributed to other diseases like flu.	36	3,92	0,806
Inadequate or lack of emergency preparedness and of response plan concerning biological risks.	36	3,61	0,934
Inappropriate measuring methods or measuring/analysing equipment for biological agents.	36	3,44	1,081

Experts' comments

Risk strongly agreed as emerging (MV > 4)

- Poor or difficult assessment of biological risks

It is emphasised that the proper assessment of biological risks is essential to prevention. According to one respondent, the assessment of biological risks is usually more properly done in laboratories and in the health care sector. However, the identification and measurement of biological agents in general are still major concerns that need to be addressed.

Risks agreed as emerging (3.25 < MV ≤ 4)

- Poor maintenance of air-conditioning and water systems

Workers involved in maintenance activities are especially exposed to legionella.

- Inadequate emergency response plan to biological risks

Pandemic plans are under development to ensure preparedness and response in case of epidemic outbreak.

- Inappropriate measuring methods or equipment for biological agents

German legislation does not require biological agents to be measured, as no occupational exposure limits (OELs) exist. As a consequence, it is often wrongly believed that there are no standards applying to biological agents. Another issue raised by one respondent is that the market for measuring equipment for biological agents is not a large one, and there is, therefore, no great interest from the producers.

5.4. Biological risks linked to social and environmental phenomena

Diagram 9. Biological risks linked to social and environmental phenomena identified in the survey

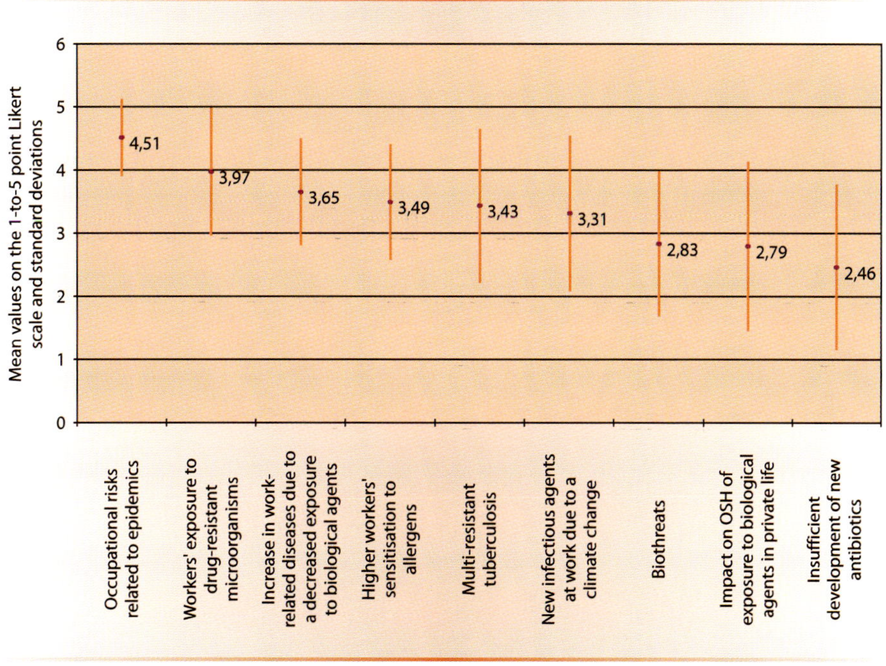

Six of the nine biological risks linked to social and environmental phenomena are agreed (3.25<MV≤4) or even strongly agreed (MV>4) to be emerging risks.

In particular, occupational risks related to global epidemics are strongly agreed as emerging risks with a good consensus among the experts (Table 6). Examples of diseases mentioned by the experts are severe acute respiratory syndrome (SARS), viral hemorrhagic fever, tuberculosis, acquired immune deficiency syndrome (AIDS), hepatitis C, hepatitis B, etc. More information on emerging or re-emerging epidemics affecting the occupational environment in the context of the changing world of work are available in the literature review '4.2.1 Occupational risks related to global epidemics'. Occupational risks in the context of pandemics have also been identified as one of the main OSH priorities in a review of various national, EU and international resources aimed at identifying future EU research needs in the field of OSH, and carried out by the Agency [47].

The risk of workers' contamination with drug-resistant microorganisms, especially in the health care sector and in the food-manufacturing industry, has been identified as emerging (see literature review '4.2.2 Workers' exposure to antimicrobial-resistant pathogens in the health care sector and livestock industry').

(47) European Agency for Safety and Health at Work, 'Priorities for occupational safety and health research in the EU-25' Luxembourg, 2005

Although proposed by a participant in the first survey round, 'insufficient development of new antibiotics' has been agreed as non-emerging (2= MV<2.75) in the subsequent survey rounds.

Table 6: Prioritised list of biological risks linked to social and environmental phenomena identified in the survey (N=number of experts answering the specific item; mean value; standard deviation)

MV>4: risk strongly agreed as emerging	2.75≤MV≤3.25: status undecided
3.25<MV≤4: risk agreed as emerging	2≤MV<2.75: risk agreed as non-emerging

NB: None of the risks were strongly agreed as non-emerging (MV<2).

Biological risks linked to social and environmental phenomena	N	Mean Value (MV)	Standard Deviation (SD)
Globalisation leading to epidemics of old and new pathogens (e.g. Severe Acute Respiratory Syndrome (SARS), avian flu, viral hemorrhagic fever, tuberculosis, Human Immunodeficiency Virus (HIV), Hepatitis C, Hepatitis B): • High density of animals in confined spaces in contact with humans leading to increasing zoonosis cases (diseases jumping the species barrier from animals to humans). • High population density and increase in business trips, tourism and immigration helping zoonoses and other infectious diseases to spread quickly worldwide. Groups particularly at risks of contamination: staff involved in producing, processing and transporting livestocks, airport staff and air crews, staff involved in border controls, policing, staff in health care sector, public transport and public services. The risk is often underestimated, which leads to a lack of preventive measures.	35	4,51	0,612
General increased use of antibiotics for human health care and for animal breeding in the food industry leading to the apparition of drug resistant pathogens (e.g., methicillin resistant staphylococcus aureus (MRSA), tubercule bacillius (TBC)). Health effects observed: increase in staff infected with MRSA in western hospitals; increasing antibiotics resistance of livestock farmers and in the population in general.	35	3,97	1,014
Decreasing exposure to biological agents — especially in developped countries, where there is a misunderstanding of hygiene — leading to a poor development of immunoregulatory pathways and to an increasing incidence of allergies, infectious diseases, arteriosclerosis, autoimmune diseases, cancers, etc.. (Studies show that the decreasing exposure to organic dusts, endotoxins from gram-negative bacteria, mycobacterial lipopeptides and fungal glucans has led to an increased morbidity especially in occupations where organic dust is to be found (livestock farmings, cotton textile industry, etc.).)	34	3,65	0,849
Environmental allergens leading to a higher sensitisation of the workforce and hence to an increase in occupational allergic diseases (atopy).	35	3,49	0,919
Multi-resistant tuberculosis coming back from Eastern Europe.	35	3,43	1.220
Climate change (warmer temperatures) may lead to the development and spread of new infectious diseases in different workplaces.	35	3,31	1,231
Increased risks of biothreats (e.g., anthrax or ricin) leading to risks of infectious diseases, poisoning and stress-related disorders.	35	2,83	1.150

Biological risks linked to social and environmental phenomena	N	Mean Value (MV)	Standard Deviation (SD)
Biological risks in private life that have a direct or indirect impact on occupational life.	34	2,79	1,343
Low interest of the pharmaceutical industry in developing new types of antibiotics leading to risks of epidemics of infectious deseases.	35	2,46	1,314

Experts' comments

Risk strongly agreed as emerging (MV > 4)

- Occupational risks related to global epidemics

There is a real risk of global epidemics of endemic diseases such as malaria, dengue fever, and of meningococcal disease and measles, etc. The following groups of workers face increased risks of contracting such diseases in their jobs: health care staff, livestock handlers, airport staff and workers involved in border controls. Air crew are also at risk because they are exposed to poorly filtered air recirculated into aircraft cabins. Additionally, drivers in public transport are at risk of coming into contact with infected people and being contaminated. With regards to the livestock industry, close contact between human beings and animals in confined spaces is a long-established practice that has contributed to influenza pandemics through antigenic shifts in the past. According to the respondents, increased global travel is another factor responsible for the increasing risk of pandemics. In this regard, the need for better information systems alerting travellers to these risks is emphasised. Generally, there is a need for more detailed data on workers' groups at risk in order to help employers to implement preventive measures, and to give policy-makers evidence of the need for more research funding.

Risks agreed as emerging (3.25 < MV ≤ 4)

- Workers' exposure to drug-resistant bacteria in the health care sector and in the food industry

The increased use of antibiotics for human health care and for animal breeding, as well as the inadequate use thereof (for example, too low dosage or a treatment not followed until completion), or the use of an antibiotic for which there is no resistance study, lead to the emergence of drug-resistant bacteria, such as multiresistant tuberculosis. However, despite supporting evidence for this phenomenon, quantitative epidemiological information is still rather weak and more research needs to be conducted in this field.

- Increase in work-related diseases due to a decreased exposure to biological agents

If increased exposure to certain biological material may lead to an increase in, for example, infectious diseases, too little exposure to biological agents can also lead to a weaker immune system and may result in an increased occurrence of (work-related) allergies and asthma. However, while issues such as global epidemics and the emergence of drug-resistant pathogens receive more public attention and research funding, the problems resulting from lower exposure to biological material are not so visible. Nevertheless, there is a need for more research into the mechanisms underlying the development of diseases such as allergies and the dose-response relationships.

- Workers' higher sensitisation to allergens

It is difficult to differentiate between an allergy resulting from occupational exposure to allergens in the workplace and an allergy resulting from exposure to environmental (i.e. ubiquitous) allergens.

- New infectious agents at work due to climate change

The average temperature rise observed in the last century has favoured the importation of new disease-vectors into Europe — such as mosquitoes transmitting malaria, or phlebotomine sand-flies which are vectors of Leishmania — and posing a risk to workers, for instance, involved in import activities. Conversely, climate change may lead to the disappearance of other diseases — although some respondents question whether climate change is actually occurring.

Undecided (2.75 ≤ MV ≤ 3.25)

- Impact on OSH of exposure to biological agents outside work

There is sometimes an overlap between occupational health and public health. While some occupational biohazards may pose a threat to public health, exposure to biological agents outside work may introduce pathogens into the workplace and turn into occupational concerns (for example, a worker who has caught an infection away from work may introduce it into the working environment and be a source of infection for his or her work colleagues).

6.

CONCLUSION

'We all swim in a single microbial sea' declared the WHO director-general referring to the severe acute respiratory syndrome (SARS) outbreak in 2003. This also includes the world of work and, indeed, two of the major concerns highlighted in this forecast, namely the occupational risks linked to global epidemics and the issue of drug-resistant organisms in the workplace, illustrate how important it is that biological risks be dealt with as a global issue. While workers are often at the front line of the risk of contamination with biological agents, at the same time, they are in a position to act as a bulwark against the spreading of epidemics. Conversely, some infection sources could be identified and controlled at earlier stages of the epidemiological chain even before they enter into the workplace and pose an occupational risk to workers. There is therefore a need to consider all collective responsibilities and means of control, both inside and outside the workplace, to tackle biological hazards appropriately. This underlines the importance of close cooperation between occupational safety and health actors and other authorities, such as public health, animal health, environmental protection, and food safety. Although these fields are outside the remit of the Agency, it is essential to stress the importance of such multi-disciplinary cooperation and coordination. Effective use and sharing of research and information is of the utmost importance.

This forecast also shows that knowledge of biohazards is still relatively scarce and that biological risks are not yet adequately managed in the workplace. While biological agents are better assessed and controlled when their use is intentional, such as in microbiological laboratories, there is a need for improvment where their presence is an unintentional consequence of the work, for instance in waste sorting or agricultural activities, and especially in SMEs. In terms of research and development needs, more reliable methods for the measurement of biological agents must be developed and standardised — even through a new European standard — in order to have harmonised, comparable quantitative exposure assessments. Reliable measurement methods and results will also enable us to understand better the complex, yet not-well-established dose-effect relationships for biological agents. Indeed, although exposure to complex mixtures of dangerous substances is a common feature to many workplaces, it is difficult to determine which biological agent accounts for which health effects, and at which exposure doses, all the more as this also depends on individual susceptibility. Exposure assessment and dose-effect relationships are key elements towards proper risk assessment. Last but not least, the still sparse state of knowledge on biological agents hinders the establishment of occupational exposure limits (OELs), which would offer helpful guidance for proper exposure assessment.

The results of this forecast, together with the three complementary forecasts on physical risks, chemical risks, and psychosocial risks, are only the first steps in a process of debate and consolidation that forms part of the work program of the Agency. In this context, they were discussed by representatives from major European OSH research institutes and from UNICE, ILO, DG Research and DG Employment, in a seminar, 'Promoting OSH research in the EU', organised by the Agency (Bilbao, 1st-2nd December 2005). During this seminar, several of the emerging risks identified in the forecasts were agreed for inclusion into a consensus list of top OSH research priorities. One aim of this list is to make these priorities more visible to policy-makers and to promote their inclusion into the seventh research framework programme (FP7). Additionnally, in June 2007, a workshop dedicated to the issue of occupational risks arising from biological agents in the workplace will bring together high-level representatives of the OSH community and from further disciplines concerned with the issue of biological risks — such as public health, animal health, food safety, environmental protection — as well as policy-makers and social partners in order to stimulate debate on the risks identified in this forecast and explore concrete ways to tackle them.

Because the world of work is constantly changing, a feasibility study for a future large-scale forecasting study is currently being undertaken, building on the experience gained through these four Delphi surveys. The future study should enable the long-term follow-up of the constant technical and societal evolution and provide a continuously up-to-date forecast on emerging OSH risks.

All the results from this work of the European Risk Observatory are available in a dedicated web feature (48), accessible from the web site of the European Agency for Safety and Health at Work (49).

(48) http://riskobservatory.osha.europa.eu/

(49) http://osha.europa.eu/OSHA

European Agency for Safety and Health at Work

EUROPEAN RISK OBSERVATORY REPORT

ANNEXES

Annex 1: Organisations contacted for the survey on emerging OSH biological risks

Country	Organisations in which experts were invited to participate	Response to at least one round
Austria	AMD-Linz	No
Austria	AUVA — Allgemeine Unfallversicherungsanstalt Austrian Social Insurance for Occupational Risks	Yes
Austria	Igeneon	Yes
Austria	Institut of Hygenie, Medical University Graz	Yes
Belgium	Arbeids-en verzekeringsgeneeskunde U.Z.	No
Belgium	Bayer CropScience	Yes
Belgium	Centre Scientifique et Technique de la Construction	No
Belgium	Faculteit Geneeskunde, Katholieke Universiteit Leuven	No
Belgium	FOD Werkgelegenheid, Arbeid en Sociaal Overleg	No
Belgium	GlaxoSmithKline Biologicals	Yes
Belgium	Heymansinstituut voor farmacologie UGent	No
Belgium	Janssen Pharmaceuticalaan 3	No
Belgium	Scientific inst. Of Public Health	No
Belgium	UCL	No
Belgium	ULB	No
Belgium	ULB Hôpital Erasme	No
Belgium	ULB Institut de Pharmacie	No
Belgium	ULG	No
Belgium	VITO	No
Bulgaria	Head of Laboratory 'Hepatitis viruses'	No
Bulgaria	Head of Microbiology Department	Yes
Cyprus	Department of Labour Inspection, Ministry of labour and Social Insurance	Yes
Czech Republic	Ministry of Health	Yes
Czech Republic	National Institute of Public Health	No
Denmark	Danish Working Environment Service National Working Environment Authority	No
Denmark	Department of Environmental and Occupational Medicine University of Aarhus	Yes
Denmark	National Institute of Occupational Health	Yes
Denmark	Working Environment Authority	Yes
Estonia	Health Care Board Occupational Health Department	Yes
Estonia	Labour Inspectorate	Yes
Estonia	Ministry of Social Affairs of Estonia	No
Finland	Ministry of Social Affairs and Health	Yes
Finland	National Public Health Institute	Yes

Country	Organisations in which experts were invited to participate	Response to at least one round
Finland	Uusimaa Regional Institute of Occupational Health	No
France	INRS	Yes
Germany	Abbott GmbH & Co. KG	Yes
Germany	Berufsgenossenschaft der chemischen Industrie	No
Germany	Berufsgenossenschaft für Fahrzeughaltungen, Hamburg	No
Germany	Berufsgenossenschaft Nahrungsmittel u. Gaststätten Mannheim	Yes
Germany	Berufsgenossenschaftliches Institut für Arbeitsschutz	Yes
Germany	BioImmunPharma GmbH Immune Biotec Pharma Consulting	Yes
Germany	City of Hamburg, Administration	Yes
Germany	Federal Institute for Occupational Safety and Health BAuA	Yes
Germany	Federal Ministry of Economics and Labour	Yes
Germany	Hannover Medical School	Yes
Germany	Hauptverband der gewerblichen Berufsgenossenschaften, Berufsgenossenschaftliche Zentrale für Sicherheit und Gesundheit — BGZ	Yes
Germany	Landesamt für Arbeitsschutz Potsdam	Yes
Germany	Landesanstalt für Arbeitsschutz, LafA, Nord-Rhein-Westfallen	Yes
Germany	Landesinstitut für Arbeitsschutz und Arbeitsmedizin, Brandenburg	Yes
Germany	Robert Koch Institut	Yes
Germany	Roche Diagnostics GmbH Penzberg	Yes
Germany	Süddeutsche Metall-Berufsgenossen-schaft	Yes
Germany	Tiefbau-Berufsgenossenschaft	Yes
Germany	University Heidelberg	No
Germany	Verband Deutscher Betriebs- und Wertsärtze (VDBW)	Yes
Hungary	National Institute of Occupational Health	No
Iceland	Administration of Occupational Safety and Health	No
Ireland	Health and Safety Authority	Yes
Italy	ISPESL	Yes
Italy	University of Padua	Yes
Latvia	Institute of Occupational and Environmental Health	No
Latvia	Ministry of Welfare	Yes
Lithuania	Kaunas University of Medicine, Department of Occupational and Environmental Medicine	No
Lithuania	Vilnius University, Medicines faculties, Public Health Institute	Yes
Malta	Occupational Health And Safety Authority	No
Netherlands	Groningen University	No
Netherlands	IRAS Institute for risk Assessment Sciences	Yes
Netherlands	Nederlands Centrum voor Beroepsziekten (Dutch centre for oocupational diseases)	Yes
Netherlands	TNO Nutrition and Food Research	Yes
Netherlands	University Maastrich	No
Netherlands	University Medical Centre Nijmegen	No

Country	Organisations in which experts were invited to participate	Response to at least one round
Netherlands	VU University Medical Center Amsterdam	No
Netherlands	Wageningen University	Yes
Norway	Det Norske Veritas	No
Poland	Central Institute for Labour Protection — National Research Institute	Yes
Romania	Bucharest Public Health Institute	Yes
Romania	Cluj-Napoca Public Health Institute	No
Romania	National Research Institue for Labour Protection	Yes
Spain	Dirección General de Trabajo. Consejería de Empleo y Desarrollo Tecnológico	Yes
Spain	INSHT- Centro Nacional de Condiciones de Trabajo	Yes
Spain	Inspección de Trabajo y Seguridad Social	No
Spain	Servicio de Prevención de riesgos laborales del Instituto de Salud Carlos III	No
Spain	Servicio de Prevención del Ministerio de Economía	No
Sweden	Arbetslivsinstitutet	Yes
Sweden	Swedish Institute for Infectious Disease	No
Sweden	Uppsala universitet, Institutionen för medicinska vetenskaper	No
Switzerland	Institut of Virology and Immunoprophylaxis	No
Switzerland	Institute for Health at Work (IST) in Lausanne	Yes
Switzerland	Institute of Virology and Immunoprophylaxis	Yes
Switzerland	Janal & Partner Biosafety Consulting	Yes
Switzerland	Novartis International AG	Yes
Switzerland	SUVA	No
Switzerland	Swiss Federal Office of Public Health	Yes
UK	Amicus-MSF	No
UK	AstraZeneca UK Ltd	Yes
UK	Dept of Pharmacy. Kings College London	No
UK	HPA/ CAMR	No
UK	HPA/ CDSC	No
UK	HSE	Yes
UK	HSL	No
UK	Imperial College London	No
UK	Medical Research Council — MRC Environmental Epidemiology Unit	Yes
UK	Novartis Horsham Research Centre	No
UK	St. Thomas's Hospital; London	No
UK	Union of Shop, Distributive and Allied Workers	No
UK	University of Manchester	Yes
UK	University of Stirling	No
UK	University of Glasgow	Yes

Annex 2: Questionnaire used for the first survey round

Expert forecast on emerging OSH* biological risks
First survey: Identification of risks

* OSH: Occupational Safety and Health.

The survey

As part of an ongoing project on emerging health and safety at work risks, the European Agency's Topic Centre Research is formulating 'expert forecasts' in a number of areas.

This survey is the first step in the production of an expert forecast in the area of emerging OSH biological risks. It aims to create a list of potential emerging OSH biological risks and their context (cause, impact on workers' health, etc.). The results will be validated in a further survey round in order to establish a degree of consensus among the experts.

'Emerging risks' — definition

For this project, an 'emerging OSH risk' is any occupational issue that is suspected to be a risk and that is both **new** and **increasing**.

By **new** we mean that:

- the issue is new and caused by new types of biological agents, new processes, new technologies, new types of workplaces, or social or organisational change; or,
- a long-standing issue is newly considered as a risk due to a change in social or public perceptions (e.g. stress, bullying); or,
- new scientific knowledge allows a longstanding issue to be identified as a risk (e.g. allergens, such as the cow hair allergen, which have existed for decades without being identified as the cause of the allergic reactions induced).

The risk is '**increasing**' if either the:

- number of hazards leading to the risk is growing, or the
- likelihood of exposure to the hazard leading to the risk is increasing, (exposure degree and/or the number of people exposed), or the
- effect of the hazard on the workers' health is getting worse.

How to complete the questionnaire

!!! Please note that the aim of this questionnaire is not to produce a detailed list of all biological agents that are (potentially) dangerous!!!

We ask you to identify up to five risks that in your opinion are emerging risks, according to the definition above, and to give some information about why you think this is the case. Consider not only new work situations, but also changing public perceptions and the development of knowledge about longstanding issues. Similarly, a risk is increasing not only when there is a higher likelihood of exposure, but also if there are new combined effects or if a different, more vulnerable, group is exposed.

Below are possible questions that you may ask yourself in order to identify emerging OSH biological risks. (The examples in parentheses only aim at illustrating the questions but are not necessary emerging risks.)

- Are there new groups or types of biological agents (e.g. SARS; avian flu; biological agents in waste treatment and processing) likely to lead to (new) occupational diseases or work related diseases?
- Is there an increased use of types of biological agents likely to provoke more diseases? (e.g. use of enzyme-producing micro-organisms leading to an increase of respiratory allergies)?
- Are there new forms of 'old' biological agents (e.g. genetically modified organisms e.g. in food processing) that may represent a health hazard although the biological agents themselves are harmless under their usual conditions of use (form and concentration)?
- Are there new methods, new technologies, new working procedures, or new types of workplaces that could lead to new problems? (e.g. the increase of nursing at home has led to a shift of the risk of infection from hospitals into households where the danger is more difficult to control; tropical diseases such as the Ebola virus being spread by tourism and other travelling (military, humanitarian) activities to foreign countries; molds in buildings; 'lifeguard lung' in indoor pools)
- Are there long-standing issues that are becoming more important in the public perception (e.g. animal allergens; other allergenic micro-organisms such as molds for horticultural or agricultural workers) and suspected to be risks?

Use as much space as necessary for your answers. There is space at the end of the questionnaire for comments.

Please send your completed questionnaire in to emmanuelle.brun@hvbg.de **before June 4th**

Thank you very much for taking part in this survey !!!

Part 1: General information:

Please fill in:

Date:			
Name:			
Country:			
Institution:			
Function:	☐ President/ Director	☐ Head of department	☐ Professor/ Lecturer
	☐ Researcher	☐ Technician	☐ Work inspector
	☐ OSH practitioner	☐ Other:	
Main activity:	☐ Research	☐ Development	
	☐ Policy/ standards development	☐ Testing/ certification	
	☐ (Law) enforcement/ promotion	☐ Research planning/ management	
	☐ Work inspection	☐ Training/ teaching	
	☐ Consulting	☐ Other:	

Do you have at least 5 years of experience in activities related to OSH biological risks?

☐ Yes ☐ No

Part 2: Emerging OSH biological risks

In your opinion, what are the emerging OSH biological risks of the next 10 years?

You may describe up to five OSH emerging biological risks in the fields below. Please do not make a detailed list of organisms but focus on groups of biological agents, on types of work processes or technologies and the biological agents involved, etc. Please note that you may include risks due to infections as well as allergenic potential or production of toxins. (See 'How to complete the questionnaire' for more details). Use as much space as necessary for your answers.

Risk 1:

Risk 2:

Risk 3:

Risk 4:

Risk 5:

N.B.: In the following questions, Risks 1 to 5 always refer to the corresponding risks you have identified in question 1.

What are the cause(s) for the risk(s)?

Is it due, for instance, to a new type of biological agent, a new work process, a new type of workplace, a modification of the working environment, a lack of qualification, an unfavourable workplace design, etc.?

Please explain:

Risk 1:

Risk 2:

Risk 3:

Risk 4:

Risk 5:

What are the health effects of the risk(s) (occupational diseases/ work related diseases/ sickness days)?

Please describe:

Risk 1:

Risk 2:

Risk 3:

Risk 4:

Risk 5:

Where is the risk to be found?

Is it, for instance, branch specific? Or specific to a type of workplace? Or to a type of work process? Please specify if relevant:

Risk 1:

Risk 2:

Risk 3:

Risk 4:

Risk 5:

Why do you think that the risk(s) is/are new?

Is it, for instance, new because:

- The group of biological agents concerned is new? The work process or the technology involved is new? The conditions of use (form, concentration, etc.) are new?
- There is a new recognised occupational disease caused by this risk?
- The public concern/discussion about this issue is rising?
- There are more and more political debates about this issue?
- There is new scientific knowledge about it?
- There have been more requests for consultation activities from employers on this issue lately?
- New research programs on this topic have been created? Etc.

Please explain:

Risk 1:

Risk 2:

Risk 3:

Risk 4:

Risk 5:

Why do you think that the risk(s) are increasing?

Is it, for instance, because of an increase of:

- The number of hazards (e.g. increase in use of a group of biological agent)?
- The intensity of exposition to this hazard (e.g. increase of the concentration of a biological agent)?
- The number of persons exposed? (If you are able to give an indication, please do so.)
- The number of occupational/ work related diseases or sick-leaves caused by this hazard? (If you are able to give an indication, please do so.)
- Or is the effect of this hazard on the workers' health getting worse? Etc.

Please explain:

Risk 1:

Risk 2:

Risk 3:

Risk 4:

Risk 5:

Could you give us references of publications/studies dealing with these suspected emerging risks?

Risk 1:

Risk 2:

Risk 3:

Risk 4:

Risk 5:

Part 3: Further information

1. **Do you know about other studies/ publications dealing with emerging risks (not limited to biological risks)? If yes, please give references:**

2. **Can you recommend national or international experts whom we should invite to participate in this survey? Please give name, organisation, address, phone number, e-mail:**

3. **Do you have any further complementary information or comments about our project in general? Any suggestions on how to improve our questionnaire?**

Thank you very much for your time and co-operation!

Annex 3: Questionnaire used for the second survey round

Survey on emerging biological risks related to occupational safety and health (OSH) — Second round

About the survey

This survey represents the second step in the Agency's expert forecast on emerging biological occupational safety and health risks. The questionnaire seeks your opinion on the issues identified by the experts in the previous round of the survey.

It is divided into four parts, each one focusing on a particular topic in the field of biological OSH.

We would like to have your opinion:

How strongly do you agree on that the following issues are emerging biological OSH risks?

Definition of 'emerging risk'

For this project, an 'emerging OSH risk' is any occupational risk that is both 'new, and 'increasing'.

By new we mean that:

- the risk is new and caused by new processes, new technologies, new types of workplaces, or social or organisational change; or,
- a long-standing issue is newly considered as a risk due to a change in social or public perceptions (e.g. stress, bullying); or,
- new scientific knowledge allows a longstanding issue to be identified as a risk (e.g. Repetitive Strain Injury (RSI), where cases have existed for decades without being identified as RSI because of a lack of scientific evidence).

The risk is 'increasing' if either the:

- number of hazards leading to the risk is growing, or the
- likelihood of exposure to the hazard leading to the risk is increasing, (exposure level and/ or the number of people exposed), or the
- effect of the hazard on the workers' health is getting worse.

Please send the questionnaire filled in to eva.flaspoeler@hvbg.de **by February 7th, 2005.**

Thank you very much for taking part in this survey!

How to complete the questionnaire

The risks identified in the first step of the survey in 2004 are categorized and listed in tables. The first column in each of the tables gives feedback on the results of the survey's first round: It shows the number of experts who considered the risk to be emerging.

If you have at least **five years of experience in the area of biological risks**, please rate each issue independently by ticking the corresponding box on a five-point scale ranging from 'Disagree' to 'Agree'.

- Tick the first box if you strongly disagree that the issue is an emerging risk.
- Tick the last box if you strongly agree that the issue is an emerging risk.
- Tick the middle box if you are undecided.

You may comment on your ratings in the column 'Comments, on the right of each issue. If you do so, please avoid unsubstantiated opinions and try to support your comments with objective arguments, e.g. research results, references to publications, statistics, etc. At the end of each part you may also add new additional possible emerging biological risks, if in your opinion some relevant emerging biological OSH risks are missing.

Moreover, you will find some space for any additional comments on the survey in general at the end of the questionnaire.

About you

All information is kept confidential within the project team and is only used for purposes of the Agency's expert forecast project.

Date:		
Name:		
Country:		
Institution:		
Function:	☐ President/ Director	☐ Head of department	☐ Professor/ Lecturer
	☐ Researcher	☐ Engineer	☐ Work inspector
	☐ Other:		
Main activity:	☐ Research	☐ Development	
	☐ Policy/ standards development	☐ Testing/ certification	
	☐ (Law) enforcement/ promotion	☐ Research planning/ management	
	☐ Work inspection	☐ Training/ teaching	
	☐ Consulting	☐ Other:	

Part 1: Biological risks linked to social and environmental phenomena

Number of experts	Emerging biological risks linked to social and environmental phenomena	Ratings	Comments
16	Globalisation leading to epidemics of old and new pathogens (e.g. Severe Acute Respiratory Syndrome (SARS), avian flu, viral hemorrhagic fever, tuberculosis, Human Immunodeficiency Virus (HIV), Hepatitis C, Hepatitis B): • High density of animals in confined spaces in contact with humans leading to increasing cases zoonoses (diseases jumping the species barrier from animals to humans). • High population density and increase in business trips, tourism and immigration helping zoonoses and other infectious diseases to widespread quickly world-wide. Groups particularly at risks of contamination: Staff involved in producing, processing and transporting livestocks, airport staff and air crews, staff involved in border controls, policing, staff in health care sector, public transport and public services. The risk is often underestimated, which leads to a lack of preventive measures.	Disagree Agree ☐ ☐ ☐ ☐ ☐	
8	General increased use of antibiotics for human health care and for animal breeding in the food-industry leading to the apparition of drug resistant pathogens (e.g., methicillin resistant staphylococcus aureus (MRSA), tubercule bacillius (TBC)). Health effects observed: Increase in staff infected with MRSA in western hospitals; increasing antibiotics resistance of livestock farmers and in the population in general.	Disagree Agree ☐ ☐ ☐ ☐ ☐	
4	Increased risks of biothreats (e.g., anthrax, ricin) leading to risks of infectious diseases, poisoning and stress-related disorders.	Disagree Agree ☐ ☐ ☐ ☐ ☐	
2	Decreasing exposure to biological agents — especially in developed countries, where there is a misunderstanding of hygiene — leading to a poor development of immunoregulatory pathways and to an increasing incidence of allergies, infectious diseases, arteriosclerosis, autoimmune diseases, cancers,etc.. (Studies show that the decreasing exposure to organic dusts, endotoxins from gram-negative bacteria, mycobacterial lipopeptides and fungal glucans has led to an increased morbidity especially in occupations where organic dust is to be found (livestock farmings, cotton textile industry, etc.)).	Disagree Agree ☐ ☐ ☐ ☐ ☐	
1	Lack of interest from the pharmaceutical industry in developing new types of antibiotics leading to risks of epidemics of infectious deseases.	Disagree Agree ☐ ☐ ☐ ☐ ☐	

Number of experts	Emerging biological risks linked to social and environmental phenomena	Ratings	Comments
1	Climate change (warmer temperatures) may lead to the development and spread of new infectious diseases in different workplaces.	Disagree Agree ☐ ☐ ☐ ☐ ☐	
1	Environmental allergens leading to a higher sensitisation of the workforce and hence to an increase in occupational allergic diseases (atopy).	Disagree Agree ☐ ☐ ☐ ☐ ☐	

Other emerging biological risks linked to social and environmental phenomena:

Part 2: Risks due to substances

Number of experts	Emerging risks due to substances	Ratings	Comments
4	Moulds in indoor workplaces due to new construction methods and materials, due to the aim of saving energy and due to the lack of maintenance: Exposure to fungal spores for office workers and especially workers involved in building restauration, leading to sensitization and allergies.	Disagree Agree ☐ ☐ ☐ ☐ ☐	
3	Endotoxins: High concentrations in various industrial settings (e.g. in workplaces exposed to organic materials (straw, wood, cotton dust), waste treatment, poultry houses, swine confinment buildings) leading to asthma, loss of lung function, etc.	Disagree Agree ☐ ☐ ☐ ☐ ☐	
1	Aflatoxin exposure of staff in food processing plants and animal feeding plants may lead to e.g. cancer.	Disagree Agree ☐ ☐ ☐ ☐ ☐	
1	Mycotoxins: Increasing risk as mycotoxins have increasing possibilities to grow, for example due to the increase of garbage quantities. Potential health effects: cancers, immune deprivations and congenital abnormalities. Groups more at risk: workers in waste treatment occupations, textile and food-processing sectors, wet work.	Disagree Agree ☐ ☐ ☐ ☐ ☐	

Number of experts	Emerging risks due to substances	Ratings	Comments
1	Pneumococcus and various infectious agents from metal fumes, the effects of which were previously unrecognised.	Disagree Agree ☐ ☐ ☐ ☐ ☐	
1	Potentially more aggressive micro-organisms and products, mainly resulting from the increasing enzyme production and medicine production.	Disagree Agree ☐ ☐ ☐ ☐ ☐	
1	Bioaerosols and chemicals, the combined effects of which have been very little studied but lead to allergies. More knowledge will help identify the real multi-factorial causes of symptoms for which mono-causal explanations have been made so far.	Disagree Agree ☐ ☐ ☐ ☐ ☐	

Other emerging risks due to substances:

Part 3: Risks due to specific workplaces and work processes

Number of experts	Emerging risks due to workplaces and work processes related to recycling and waste handling	Ratings	Comments
7	Biohazards in waste treatment plants (e.g. selective sorting, manufacture of compost) leading to allergies, infectious diseases (bacteria, viruses), toxinic diseases (endotoxins, mycotoxins) and cancers (oncogens). Especially in composting facilities, where there are a wide variety of microorganisms present at the different stages of the composting process, the risks are not completely identified yet.	Disagree Agree ☐ ☐ ☐ ☐ ☐	
2	Increased number of water treatment plants implying a larger number of workers exposed to risks of allergies, infectious diseases (bacteria, viruses), toxinic diseases (endotoxins, mycotoxins) and cancers (oncogens).	Disagree Agree ☐ ☐ ☐ ☐ ☐	
1	Handling and processing of clinical waste.	Disagree Agree ☐ ☐ ☐ ☐ ☐	

Number of experts	Emerging risks due to workplaces and work processes related to recycling and waste handling	Ratings	Comments
1	Increased need for renovation of old degrading sewage systems and drain pipes in Europe, which are sources of many infectious agents (hepatitis A, endotoxin from gram negative bacteria).	Disagree Agree ☐ ☐ ☐ ☐ ☐	

Number of experts	Emerging risks due to workplaces and work processes related to health care and service sectors	Ratings	Comments
5	Increase in infections with hepatitis B, C, HIV in health care sector, police and prison staff.	Disagree Agree ☐ ☐ ☐ ☐ ☐	
1	Increase of nursing at home — because of pressure on medical budgets — leading to exposure of (less well trained self-employed) medical staff to infectious micro-organisms as the environmental working conditions are not controlled as well as in hospitals.	Disagree Agree ☐ ☐ ☐ ☐ ☐	
1	Manufacture and applications of viral constructs for gene therapy, which involve more staff in the manufacturing and health care sectors, with following safety issues: safety of the rDNA construct, safety of the viral vector, pathogenicity, recombination events with viral sequences in the host, safety of packaging cell line, and regulation of gene expression of the rDNa product.	Disagree Agree ☐ ☐ ☐ ☐ ☐	

Number of experts	Emerging risks due to workplaces and work processes related to laboratory and research work	Ratings	Comments
1	Research work on the pathogenicity of aerosol transmitting agents such as tubercolosis and SARS leading to increasing transmission to research workers.	Disagree Agree ☐ ☐ ☐ ☐ ☐	
1	Increasing amount of research work on biosafety level 4 (BSL4) agents in order to determine their pathogenicity, likely to lead to contamination of laboratory workers.	Disagree Agree ☐ ☐ ☐ ☐ ☐	
1	Exposure to different micro-organisms in laboratory workplaces.	Disagree Agree ☐ ☐ ☐ ☐ ☐	
1	Vaccines against pandemic flu: Potential for evolutionary drift to produce a novel strain with an antigenic profile for which there is no background immunity (e.g. reassortment between a circulating flu virus and a H5 antigen).	Disagree Agree ☐ ☐ ☐ ☐ ☐	
1	Large-scale production of vaccines against pandemic flu: Conflict in requirements between positive pressure from safety regulations (to work under controlled safety conditions) and negative pressure from public health protection (to produce the vaccines as fast as possible).	Disagree Agree ☐ ☐ ☐ ☐ ☐	

Number of experts	Emerging risks due to workplaces and work processes related to the food industry	Ratings	Comments
1	Biotechnologies, involving new substances in occupational settings (e.g. food production).	Disagree Agree ☐ ☐ ☐ ☐ ☐	
1	Use of enzymes under new conditions in the food and detergent sectors (wider and more concentrated applications) leading to respiratory and dermal allergies.	Disagree Agree ☐ ☐ ☐ ☐ ☐	

Number of experts	Emerging risks due to workplaces and work processes related to agriculture	Ratings	Comments
1	Biological pest control in green houses leading to allergies.	Disagree Agree ☐ ☐ ☐ ☐ ☐	
1	Exposure to flavivirus in forestry occupations leading to encephalitis.	Disagree Agree ☐ ☐ ☐ ☐ ☐	
1	Exposure to clostridium tetani, potentially leading to death, in the agriculture sector or leather and fur processing occupations.	Disagree Agree ☐ ☐ ☐ ☐ ☐	

Other emerging risks due to specific workplaces and work processes:

Part 4: Risks due to risk management and handling

Number of experts	Emerging risks due to risk management and handling	Ratings	Comments
4	Poor maintenance of air-conditioning (whose use is increasing) and water systems (e.g. legionella, aspergilosis in hospitals). New knowledge about the presence of legionella will help the correct diagnosis of symptoms so far wrongly attributed to other diseases like flu.	Disagree Agree ☐ ☐ ☐ ☐ ☐	
1	Inadequate training, poor knowledge of OSH or even poor basic awareness of risks of local authorities staff (e.g sewage, excavations, waste collection, etc.).	Disagree Agree ☐ ☐ ☐ ☐ ☐	
1	Lack of information on biological risks in different workplaces (e.g. office workplaces, agriculture).	Disagree Agree ☐ ☐ ☐ ☐ ☐	
1	Inappropriate measuring methods or measuring/analysing equipment.	Disagree Agree ☐ ☐ ☐ ☐ ☐	
1	Poor or difficult assessment of biological risks.	Disagree Agree ☐ ☐ ☐ ☐ ☐	

Other emerging risks due to risk management and handling:

Further Comments

Other emerging biological risks not fitting in any of the categories above:

Do you know about other studies/ publications dealing with emerging biological OSH risks? If so, please give references:

Do you have any comments about this project or about this questionnaire? If so, please comment:

Thank you very much for your time and co-operation!

Annex 4: Questionnaire used for the third survey round

Survey on emerging OSH biological risks — 3rd round

About the survey

This survey represents the final step in the Agency's expert forecast on emerging biological occupational safety and health risks. The questionnaire seeks your opinion on the issues identified by the experts in the previous two survey rounds and is divided into four parts.

We would like to have your opinion:

Which of these issues are really emerging biological OSH risks?

Definition of 'emerging risk'

For this project, an 'emerging OSH risk' is any occupational risk that is both 'new' and 'increasing'.

By 'new' we mean that:

- the risk is new and caused by new processes, new technologies, new types of workplaces, or social or organisational change; or,
- a long-standing issue is newly considered as a risk due to a change in social or public perceptions (e.g. stress, bullying); or,
- new scientific knowledge allows a longstanding issue to be identified as a risk (e.g. Repetitive Strain Injury (RSI), where cases have existed for decades without being identified as RSI because of a lack of scientific evidence).

The risk is 'increasing' if either the:

- number of hazards leading to the risk is growing, or the
- likelihood of exposure to the hazard leading to the risk is increasing, (exposure level and/or the number of people exposed), or the
- effect of the hazard on the workers' health is getting worse.

How to complete the questionnaire

For each of the parts, please ONLY reply if you have <u>at least five years of experience</u> in the area concerned.

Please rate each issue independently by ticking the corresponding box on a five-point scale ranging from 'Disagree' to 'Agree'.

- Tick the first box if you strongly disagree that the issue is an emerging risk;
- Tick the last box if you strongly agree that the issue is an emerging risk;
- Tick the middle box if you are undecided.

We have left a space at the end for your comments.

Please send the questionnaire to eva.flaspoeler@hvbg.de **by September, 27th.**

Thank you very much for taking part in this survey!!!

About you

(Information is kept confidential within the project team and is used only for the purposes of the Agency's expert forecast project)

Date:			
Name:			
Country:			
Institution:			
Function:	☐ President/ Director	☐ Head of department	☐ Professor/ Lecturer
	☐ Researcher	☐ Engineer	☐ Work inspector
	☐ Other:		
Main activity:	☐ Research	☐ Development	
	☐ Policy/ standards development	☐ Testing/ certification	
	☐ (Law) enforcement/ promotion	☐ Research planning/ management	
	☐ Work inspection	☐ Training/ teaching	
	☐ Consulting	☐ Other:	

Do you have at least 5 years of experience in activities related to OSH biological risks?
☐ Yes ☐ No

Part 1: Biological risks linked to social and environmental phenomena

	Disagree / Agree / Comments
Globalisation leading to epidemics of old and new pathogens (e.g. Severe Acute Respiratory Syndrome (SARS), avian flu, viral hemorrhagic fever, tuberculosis, Human Immunodeficiency Virus (HIV), Hepatitis C, Hepatitis B): • High density of animals in confined spaces in contact with humans leading to increasing zoonosis cases (diseases jumping the species barrier from animals to humans). • High population density and increase in business trips, tourism and immigration helping zoonoses and other infectious diseases to widespread quickly world-wide. Groups particularly at risks of contamination: Staff involved in producing, processing and transporting livestocks, airport staff and air crews, staff involved in border controls, policing, staff in health care sector, public transport and public services. The risk is often underestimated, which leads to a lack of preventive measures.	☐ ☐ ☐ ☐ ☐
General increased use of antibiotics for human health care and for animal breeding in the food-industry leading to the apparition of drug resistant pathogens (e.g., methicillin resistant staphylococcus aureus (MRSA), tubercule bacillius (TBC)). Health effects observed: Increase in staff infected with MRSA in western hospitals; increasing antibiotics resistance of livestock farmers and in the population in general.	☐ ☐ ☐ ☐ ☐

Decreasing exposure to biological agents — especially in developed countries, where there is a misunderstanding of hygiene — leading to a poor development of immunoregulatory pathways and to an increasing incidence of allergies, infectious diseases, arteriosclerosis, autoimmune diseases, cancers,etc.. (Studies show that the decreasing exposure to organic dusts, endotoxins from gram-negative bacteria, mycobacterial lipopeptides and fungal glucans has lead to an increased morbidity especially in occupations where organic dust is to be found (livestock farmings, cotton textile industry, etc.)).	Disagree Agree Comments ☐ ☐ ☐ ☐ ☐
Environmental allergens leading to a higher sensitisation of the workforce and hence to an increase in occupational allergic diseases (atopy).	Disagree Agree Comments ☐ ☐ ☐ ☐ ☐
Increased risks of biothreats (e.g., anthrax, ricin) leading to risks of infectious diseases, poisoning and stress-related disorders.	Disagree Agree Comments ☐ ☐ ☐ ☐ ☐
Climate change (warmer temperatures) may lead to the development and spread of new infectious diseases in different workplaces.	Disagree Agree Comments ☐ ☐ ☐ ☐ ☐
Low interest of the pharmaceutical industry in developing new types of antibiotics leading to risks of epidemics of infectious deseases.	Disagree Agree Comments ☐ ☐ ☐ ☐ ☐
Multi-resistant tuberculosis coming back from Eastern Europe.	Disagree Agree Comments ☐ ☐ ☐ ☐ ☐
Biological risks in private life which have a direct or indirect impact on occupational life.	Disagree Agree Comments ☐ ☐ ☐ ☐ ☐

Part 2: Biological risks due to substances

Endotoxins: High concentrations in various industrial settings (e.g. in workplaces exposed to organic materials (straw, wood, cotton dust), waste treatment, poultry houses, swine confinment buildings) leading to asthma, loss of lung function, etc.	Disagree Agree Comments ☐ ☐ ☐ ☐ ☐
Moulds in indoor workplaces due to new construction methods and materials, due to the aim of saving energy and due to the lack of maintenance: Exposure to fungal spores for office workers and especially workers involved in building restauration, leading to sensitization and allergies.	Disagree Agree Comments ☐ ☐ ☐ ☐ ☐
Bioaerosols and chemicals, the combined effects of which have been very little studied but lead to allergies. More knowledge will help identify the real multi-factorial causes of symptoms for which mono-causal explanations have been made so far.	Disagree Agree Comments ☐ ☐ ☐ ☐ ☐
Mycotoxins: Increasing risk as mycotoxins have increasing possibilities to grow, for example due to the increase of garbage quantities. Potential health effects: cancers, immune deprivations and congenital abnormalities. Groups more at risk: workers in waste treatment occupations, textile and food-processing sectors, wet work.	Disagree Agree Comments ☐ ☐ ☐ ☐ ☐

Aflatoxin exposure of staff in food processing plants and animal feeding plants may lead to e.g. cancer.	Disagree Agree Comments ☐ ☐ ☐ ☐ ☐
Pneumococcus and various infectious agents from metal fumes, the effects of which were previously unrecognised.	Disagree Agree Comments ☐ ☐ ☐ ☐ ☐
Potentially more aggressive micro-organisms and products, mainly resulting from the increasing enzyme and medicine.	Disagree Agree Comments ☐ ☐ ☐ ☐ ☐
Impact of biofilms on public health, e.g. on water and air systems.	Disagree Agree Comments ☐ ☐ ☐ ☐ ☐

Part 3: Biological risks due to specific workplaces and work processes

Part 3.1: Emerging risks due to workplaces and work processes related to recycling and waste handling

Biohazards in waste treatment plants (e.g. selective sorting, manufacture of compost) leading to allergies, infectious diseases (bacteria, viruses), toxinic diseases (endotoxins, mycotoxins) and cancers (oncogens). Especially in composting facilities, where there are a wide variety of microorganisms present at the different stages of the composting process, the risks are not completely identified yet.	Disagree Agree Comments ☐ ☐ ☐ ☐ ☐
Increased number of water treatment plants implying a larger number of workers exposed to risks of allergies, infectious diseases (bacteria, viruses), toxinic diseases (endotoxins, mycotoxins) and cancers (oncogens).	Disagree Agree Comments ☐ ☐ ☐ ☐ ☐
Handling and processing of clinical waste.	Disagree Agree Comments ☐ ☐ ☐ ☐ ☐
Increased need in renovation of old degrading sewage systems and drain pipes in Europe, which are sources of many infectious agents (hepatitis A, endotoxin from gram negative bacteria).	Disagree Agree Comments ☐ ☐ ☐ ☐ ☐

Part 3.2: Emerging risks due to workplaces and work processes related to health care and service sectors

Increase in infections with hepatitis B, C, HIV in health care sector, police and prison staff as well as in other workplaces.	Disagree Agree Comments ☐ ☐ ☐ ☐ ☐
Manufacture and applications of viral constructs for gene therapy, which involve more staff in the manufacturing and health care sectors, with following safety issues: safety of the rDNA construct, safety of the viral vector, pathogenicity, recombination events with viral sequences in the host, safety of packaging cell line, and regulation of gene expression of the rDNa product.	Disagree Agree Comments ☐ ☐ ☐ ☐ ☐
Increase of nursing at home — because of pressure on medical budgets — leading to exposure of (less well trained self-employed) medical staff to infectious micro-organisms as the environmental working conditions are not controlled as well as in hospitals.	Disagree Agree Comments ☐ ☐ ☐ ☐ ☐

Part 3.3: Emerging risks due to workplaces and work processes related to laboratory and research work

Exposure to different micro-organisms in laboratory workplaces.	Disagree ☐ ☐ ☐ ☐ ☐	Agree	Comments
Increasing amount of research work on biosafety level 4 (BSL4) agents in order to determine their pathogenicity, likely to lead to contamination of laboratory workers.	Disagree ☐ ☐ ☐ ☐ ☐	Agree	Comments
Research work on the pathogenicity of aerosol transmitting agents such as tubercolosis and SARS leading to increasing transmission to research workers.	Disagree ☐ ☐ ☐ ☐ ☐	Agree	Comments
Vaccines against pandemic flu: Potential for evolutionary drift to produce a novel strain with an antigenic profile for which there is no background immunity (e.g. reassortment between a circulating flu virus and a H5 antigen).	Disagree ☐ ☐ ☐ ☐ ☐	Agree	Comments
Large-scale production of vaccines against pandemic flu: Conflict in requirements between positive pressure from safety regulations (to work under controlled safety conditions) and negative pressure from public health protection (to produce the vaccines as fast as possible).	Disagree ☐ ☐ ☐ ☐ ☐	Agree	Comments
Laboratory acquired infections, regardless of biosafety level.	Disagree ☐ ☐ ☐ ☐ ☐	Agree	Comments
Increasing numbers of laboratories handling highly dangerous pathogens (because of potential biothreats, epidemic strains) while not always up-to-date from a biosafety point of view (relatively low level of control by authorities especially in academic settings).	Disagree ☐ ☐ ☐ ☐ ☐	Agree	Comments

Part 3.4: Emerging risks due to workplaces and work processes related to the food industry

Use of enzymes under new conditions in the food and detergent sectors (wider and more concentrated applications) leading to respiratory and dermal allergies.	Disagree ☐ ☐ ☐ ☐ ☐	Agree	Comments
Biotechnologies, involving new substances in occupational settings (e.g. food production).	Disagree ☐ ☐ ☐ ☐ ☐	Agree	Comments

Part 3.5: Emerging risks due to workplaces and work processes related to agriculture

Biological pest control in green houses leading to allergies.	Disagree ☐ ☐ ☐ ☐ ☐	Agree	Comments
Exposure to flavivirus in forestry occupations leading to encephalitis.	Disagree ☐ ☐ ☐ ☐ ☐	Agree	Comments
Exposure to clostridium tetani, potentially leading to death, in the agriculture sector or leather and fur processing occupations.	Disagree ☐ ☐ ☐ ☐ ☐	Agree	Comments

Part 4: Biological risks due to risk management and handling

Poor maintenance of air-conditioning (whose use is increasing) and water systems (e.g. legionella, aspergilosis in hospitals). New knowledge about the presence of legionella will help the correct diagnosis of symptoms so far wrongly attributed to other diseases like flu.	Disagree ☐ ☐ ☐ ☐ ☐ Agree	Comments	
Poor or difficult assessment of biological risks.	Disagree ☐ ☐ ☐ ☐ ☐ Agree	Comments	
Lack of information on biological risks in different workplaces (e.g. office workplaces, agriculture).	Disagree ☐ ☐ ☐ ☐ ☐ Agree	Comments	
Inadequate training, poor knowledge of OSH or even poor basic awareness of risks of local authorities staff (e.g sewage, excavations, waste collection, etc.).	Disagree ☐ ☐ ☐ ☐ ☐ Agree	Comments	
Inappropriate measuring methods or measuring/analysing equipment.	Disagree ☐ ☐ ☐ ☐ ☐ Agree	Comments	
Inadequate or lack of emergency preparedness and/or response plan concerning biological risks.	Disagree ☐ ☐ ☐ ☐ ☐ Agree	Comments	

Further information

Do you know about other studies/ publications dealing with emerging OSH risks? If so, please give references:

Do you have any comments about this project or about this questionnaire?

Thank you very much for your time and co-operation!

Annex 5: References used in the literature reviews

[1] World Health Organisation (WHO), *Revision of the International Health Regulations*, Agenda item 13.1, Fifty-eighth world health assembly — WHA 58.3, 23rd May 2005, http://www.who.int/csr/ihr/WHA58_3-en.pdf

[2] World Health Organisation (WHO), *The International Health Regulations*, 2005, http://www.who.int/csr/ihr/One_pager_update_new.pdf

[3] World Health Organisation (WHO), Food and Agriculture Organisation, World Organisation for Animal Health, *Report of the WHO/FAO/OIE joint consultation on emerging zoonotic diseases*, 3rd-5th May 2004 — Geneva, Switzerland, WHO/CDS/CPE/ZFK/2004.9, http://whqlibdoc.who.int/hq/2004/WHO_CDS_CPE_ZFK_2004.9.pdf

[4] Brown, D., 'Emerging zoonoses and pathogens of public health significance — an overview', *Rev. Sci. Tech.* Vol. 23, No. 2, pp. 435–42, 2004.

[5] Thiermann, A., 'Emerging diseases and implications for global trade', *Rev. Sci. Tech.* Vol. 23, No. 2, 2004, pp. 701–7. http://www.oie.int/eng/publicat/rt/2302/PDF/701-708thierman.pdf

[6] World Health Organisation (WHO) *Ten things you need to know about pandemic influenza*, 5th Jan 2006.

[7] European Centre for Disease prevention and Control — ECDC, *Interim ECDC Risk Assessment — The Public Health Risk from Highly Pathogenic Avian Influenza Viruses Emerging in Europe with Specific Reference to type A/H5N1*. http://www.ecdc.eu.int/avian_influenza/H5N1_European_Risk_Assessment_ECDC_051019.pdf

[8] Bosman, A. et al., *Vogelpest Epidemie 2003, Gevolgen voor de volksgezondheid (Avian flu epidemic 2003, public health consequences)*. Rijksinstituut voor Volksgezondheid en Milieu (RIVM). Report 630940001, 2004, http://www.rivm.nl/bibliotheek/rapporten/630940001.pdf

[9] Food and Agriculture Organisation of the United Nations — FAO, *High bird flu risk in Africa after outbreak in Nigeria*, 8th Feb 2006, http://www.fao.org/newsroom/en/news/2006/1000226/index.htm

[10] European Centre for Disease Prevention and Control — ECDC, http://www.ecdc.eu.int/index.php

[11] Wetlands International, *Wild birds unlikely culprits in Nigeria bird flu outbreak*, http://www.wetlands.org/news.aspx?ID=bb47207c-3109-4d86-826a-d438ada02b83, Press release 8th Feb 2006

[12] Wetlands International, *History of outbreaks*, http://www.wetlands.org/articlemenu.aspx?id=8d6e360f-b67f-47f5-a299-cbb37532f5ac

[13] Van Borm, S., Thomas, I., Hanquet, H., Lambrecht, B., et al., 'Highly pathogenic H5N1 influenza virus in smuggled Thai eagles, Belgium', *Emerging infectious diseases* Vol. 11, May 2005, CDC, pp. 702–5, www.cdc.gov/eid

[14] Agence française de sécurité sanitaire des aliments: 'Avis de l'Agence française de sécurité sanitaire des aliments relatif à l'évaluation du risqué d'introduction sur le territoire national et les DOM-TOM par l'avifaune de virus influenza hautement pathogène au regard du récent foyer russe de Toula', *Afssa — Saisine n° 2005-SA-0323*, October 2005, http://www.afssa.fr/ftp/afssa/32376-32377.pdf

[15] Influenza team, European Centre for Disease Surveillance and Control Stockholm, Sweden, 'World avian influenza update: H5N1 could become endemic in Africa', *Eurosurveillance* Surveillance report Vol. 11, No. 6, 22nd June 2006, http://www.eurosurveillance.org/ew/2006/060622.asp#3

[16] Servas, V., Mailles, A., Neau, D., Castor, C., et al., 'An imported case of canine rabies in Aquitaine: Investigation and management of the contacts at risk', *Eurosurveillance* Vol. 10, No. 11, pp. 222–5, 2005, http://www.eurosurveillance.org/em/v10n11/1011–225.asp

[17] Cunha, B. E., 'Monkeypox in the United States: an occupational health look at the first cases', *AAOI IN J.* Vol. 52, No. 4, 2004, pp. 164–8.

[18] Guarner, J., Johnson, B. J., Paddock, C. D., Shieh, W. J., et al., 'Monkeypox transmission and pathogenesis in prairie dogs', *Emerging infectious diseases* Vol. 10, No. 3, 2004 pp. 426–31.

[19] Reed, K. D., Melski, J. W., Graham, M. B., Regnery, R. L., et al., 'The detection of monkeypox in humans in the Western Hemisphere', *N Engl J Med* Vol. 350, No. 4, 2004, pp. 342–50.

[20] Cunningham, A. A., 'A walk on the wild side — emerging wildlife diseases', *BMJ* No. 331, 2005, pp. 1214–5.

[21] World Health Organisation (WHO), Department of Communicable Disease Surveillance and Response, *WHO guidelines for the global surveillance of severe acute respiratory syndrome (SARS)*. Updated recommendations October 2004. Epidemic Alert and Response, WHO/CDS/CSR/ARO/2004.1, http://www.who.int/csr/resources/publications/WHO_CDS_CSR_ARO_2004_1.pdf

[22] Canadian Labour Congress, Department of Health, Safety and Environment, 'The prevention and control of communicable diseases in the workplace', strategy paper, Dec 2005, http://canadianlabour.ca/updir/PrevnContrlCommunicDis.pdf

[23] Varia, M., Wilson, S., Sarwal, S., McGeer, A., et al., 'Investigation of a nosocomial outbreak of severe acute respiratory syndrome (SARS) in Toronto, Canada', *CMAJ* Vol. 169, No. 4, 2003, pp. 285–92, http://www.cmaj.ca/cgi/content/full/169/4/285

[24] Wendong Li et al., 'Bats are natural reservoirs of SARS-like coronaviruses', *Science* No. 310(5748), 28th October 2005, pp. 676–9, http://www.sciencemag.org/cgi/content/abstract/sci;310/5748/676

[25] Crowcroft, N. S., Morgan, D., Brown, D., 'Viral haemorrhagic fevers in Europe — effective control requires a co-ordinated response'. *Eurosurveillance* Vol. 7, No. 3, 2002, pp. 31–2.

[26] Arthur, R. R., 'Ebola in Africa — discoveries in the past decade', *Eurosurveillance* Vol. 7, No. 3, 2002, pp. 33–6.

[27] Wirtz, A., Niedrig, M., Fock, R., 'Management of patients in Germany with suspected viral haemorrhagic fever and other potentially lethal contagious infections', *Eurosurveillance* Vol. 7, No. 3, 2002, pp. 36–42.

[28] Hugonnet, S., Sax, H., Pittet, D., 'Management of viral haemorrhagic fevers in Switzerland', *Eurosurveillance* Vol. 7, No. 3, 2002, pp. 42–4.

[29] Crowcroft, N., Brown, D., Gopal, R., Morgan, D., 'Current management of patients with viral haemorrhagic fevers in the United Kingdom', *Eurosurveillance* Vol. 7, No. 3, pp. 44–8.

[30] Swaan, C. M., van den Broek, P. J., Wijnands, S., van Steebergen, J. E., 'Management of viral haemorrhagic fever in the Netherlands', *Eurosurveillance* Vol. 7, No. 3, 2002, pp. 48–50.

[31] Crowcroft, N. S., 'Management of Lassa fever in European countries', *Eurosurveillance* Vol. 7, No. 3, 2002, pp. 50–52.

[32] 'Investigation autour d'un cas importé de fièvre hémorragique Crimée — Congo en France', *Bull Epidémiol Hebd.* Vol. 16, Nov 2004, pp. 61–2.

[33] 'Centers for disease control and prevention — Imported Lassa fever', *JAMA* Vol. 292, No. 33, 2004, pp. 2828–30.

[34] Rodhain, F., 'Le rôle joué par l'urbanisation et les transports dans l'évolution des maladies à vecteurs', *Mondes Cult.* Vol. 51, 1991, pp. 130–52.

[35] Mouchet, J., Giacomono, T., Julvez, J., 'La diffusion anthropique des arthropodes vecteurs de maladie dans le mond.' *Santé* Vol. 5, No. 5, 1995, pp. 293–8.

[36] Rodhain, F., 'Problèmes posés par l'expansion d'Aedes albopictus', *Bull Soc Pathol Exot.* Vol. 89, No. 2, 1996, pp. 137–40, discussion 140–41.

[37] World Health Organisation (WHO), *Dengue/dengue haemorrhagic fever,* http://www.who.int/csr/disease/dengue/en/index.html

[38] Malard, S., Schaffner, F., Lebâcle, C., 'La dengue: un problème de santé publique lié à des activités professionnelles — Lutte en entreprise contre l'introduction d'un vecteur', *Documents pour le médecin du travail*, Vol. 94, 2003, pp. 151–60.

[39] Schaffner, F., Karch, S., 'Première observation d'Aedes albopictus' (Skuse, 1984) en France métropolitaine', *C R Acad Sci III.* Vol. 323, No. 4, 2000, pp. 373–5.

[40] Le Bâcle, C., Malard, S., Schaffner, F., 'Globalised trade and the associated infectious risks — what form of regulation is required?' Communication at the 2005 World Congress on Safety and Health at Work, Orlando, USA, Sept 2005.

[41] Linthicum, J. K., Kramer, V., 'The Surveillance Team, Update on Aedes albopictus infestations in California'. *Vector Ecology Newsletter* Vol. 33, No. 1, 2002, pp. 8–10.

[42] World Health Organisation (WHO), *Marburg haemorrhagic fever — fact sheet,* 31 March 2005, http://www.who.int/csr/disease/marburg/factsheet/en/index.html

[43] Exotic Diseases Resources Associates (EDRA), *Marburg,* http://www.coppettswood.demon.co.uk/marburg.htm

[44] Buxton Bridges, C. et al., 'Risk of Influenza A (H5N1) Infection among Health Care Workers Exposed to Patients with Influenza A (H5N1), Hong Kong', *The Journal of Infectious Diseases* Vol. 181, 2000, pp. 344–8, http://www.journals.uchicago.edu/cgi-bin/resolve?id=doi:10.1086/315213

[45] Institut National de Recherche et de Sécurité — INRS. Dossier, 'Grippe aviaire: Risques professionals et prevention', http://www.inrs.fr/inrs-pub/inrs01.nsf/IntranetObject-accesParIntranetID/OM:Document:A976A3C0D79DB979C1257131004F822C/$FILE/Visu.html (Last up-date: 14th Mar 2006)

[46] Haamann, F., ‚Vogelgrippe/Klassische Geflügelpest — Informationen zu Infektionsgefahren und Behandlungsmethoden. Berufsgenossenschaft für Gesundheitsdienst und Wohlfahrtspflege', 24th Feb 2006, http://www.bgw-online.de/internet/portal/group/internetuser/page/default.psml?path=/Inhalt/OnlineInhalt/Medientypen/Fachartikel/Vogelgrippe.html

[47] International Union of Food, Agricultural, Hotel, Restaurant, Catering, Tobacco and Allied Workers' Associations — IUF, 'Avian Influenza (H5N1) and the Food Chain: The link between workers' rights, working conditions, food safety and public health', 7th Mar 2006, http://www.iuf.org/cgi-bin/dbman/db.cgi?db=default&uid=default&ID=3175&view_records=1&ww=1&en=1

[48] The Writing Committee of the World Health Organisation (WHO), 'Consultation on Human Influenza A/H5, Avian Influenza A (H5N1) Infection in Humans', *N Engl J Med* Vol. 353, No. 13, 29th Sept 2005, pp. 1374–85, http://content.nejm.org/cgi/content/full/353/13/1374

[49] Beschluss des Ausschusses für Biologische Arbeitsstoffe (ABAS), Empfehlung spezieller Maßnahmen zum Schutz der Beschäftigten vor Infektionen durch hochpathogene aviäre Influenzaviren (Klassische Geflügelpest, Vogelgrippe)', *Beschluss 608*, http://www.baua.de/nn_12420/de/Themen-von-A-Z/Biologische-Arbeitsstoffe/Ausschuss_20f_C3_BCr_20Biologische_20Arbeitsstoffe_20-_20ABAS/Informationen_20aus_20dem_20ABAS/Aktuelle_20Informationen/Beschluss608-Februar2006.pdf, Feb 2006

[50] Advisory Committee on Dangerous Pathogens, *Biological agents: Managing the risks in laboratories and healthcare premises,* May 2005, Health and Safety Executive — HSE, http://www.hse.gov.uk/biosafety/biologagents.pdf

[51] Thanawongnuwech, R., Amonsin, A., Tantilertcharoen, R., Damrongwatanapokin, S., Theamboonlers, A., Payungporn, S., et al., 'Probable tiger-to-tiger transmission of avian influenza H5N1', CDC. *Emerging infectious diseases* Vol. 11, No. 5, May 2005, http://www.cdc.gov/ncidod/EID/vol11no05/05-0007.htm

[52] World Health Organisation (WHO), *Tuberculosis and air travel — Guidelines for prevention and control* (2nd edn). WHO/HTM/TB/2006.363. 2006. ISBN 978 92 4 154698 0. http://whqlibdoc.who.int/hq/2006/WHO_HTM_TB_2006.363_eng.pdf

[53] Health and Safety Executive (HSE), *BSE and carcass disposal*, October 1996, http://www.hse.gov.uk/pubns/indg85.htm

[54] Berufsgenossenschaft Druck und Papierverarbeitung, ‚Vogelgrippe: Schutzmaßnahmen für Journalisten', http://www.bgdp.de/pages/aktuelles/Info_Vogelgrippe_fuer_Journalisten.pdf

[55] Bae, H.-G., Drosten, C., Emmerich, P., Hantson, P., et al., 'Analysis of two imported cases of yellow fever infection from Ivory Coast and the Gambia to Germany and Belgium'. *Journal of Clinical Virology* Vol. 33, 2005, pp. 274–80.

[56] Brisabois, A., Fremy, S., Gignard, A., Moury, F., 'Le paludisme des aéroports, un problème de santé publique', *Bull Epidémiol Hebd.* Vol. 29, 1996, http://www.invs.sante.fr/beh/1996/9629/index.html

[57] Marie, J. J. L., Breton, D., Polveche, Y., Davoust, B., Darre, E., Couvreur, P., 'Prévention de l'introduction d'agents biologiques en métropole depuis un théâtre d'opérations', *Méd Armées.* Vol. 33, No. 1, 2005, pp. 47–56.

[58] Centers for Disease Control and Prevention (CDC), 'Update: outbreak of severe acute respiratory syndrome — worldwide'. *MMWR Morbid Mortal Wkly Rep* Vol. 52, No. 13, 4 Apr 2003, pp. 241–8.

[59] Poutanen, S., Low, D., Henry, B., Finkelstein, S., Rose, D., Green, K., et al., 'Identification of severe acute respiratory syndrome in Canada', *N Engl J Med* Vol. 348:1995–2005, No. 20, 15th May 2003.
http://content.nejm.org/cgi/content/abstract/348/20/1995?ijkey=ab2698e206357d131c93c9740dc58f2da0714e62&keytype2=tf_ipsecsha

[60] Booth, C., Matukas, L., Tomlinson, G., Rachlis, A., Rose, D., Dwosh, H, et al., 'Clinical features and short-term outcomes of 144 patients with SARS in the Greater Toronto Area', *JAMA* Vol. 289, No. 21, 2003, pp. 2801–9, http://jama.ama-assn.org/cgi/content/abstract/289/21/2801?ijkey=9e8e35d2baeaa626fbbb15118df3fc431cd4b23a&keytype2=tf_ipsecsha

[61] Berufsgenossenschaft der Feinmechanik und Elektrotechnik, ‚Informationen zur Vogelgrippe', 13th Mar 2006,
http://www.bgfe.de/aktuell/ap_informationen_zur_vogelgrippe.htm

[62] Buxton Bridges, C., Lim, W., Hu-Primmer, J., et al., 'Risk of Influenza A (H5N1) Infection among Poultry Workers, Hong Kong, 1997–1998', *The Journal of Infectious Diseases* Vol. 185, 2002, pp. 1005–10, http://www.journals.uchicago.edu/cgi-bin/resolve?id=doi:10.1086/340044&erFrom=-2145724920240969602Guest

[63] Rial-González, E., Copsey, S., Paoli, P., Schneider, E., *Priorities for occupational safety and health research in the EU-25,* 2005, 2005European Agency for Safety and Health at Work, ISBN 92-9191-168-2,
http://osha.eu.int/publications/reports/6805648/full_publication_en.pdf

[64] Commission of the European Communities, *Communication from the Commission to the Council, the European Parliament, the European Economic and Social Committee and the Committee of the regions on strengthening coordination on generic preparedness planning for public health emergencies at EU level.* Brussels, 28.11.2005. COM(2005) 605 final. http://eur-lex.europa.eu/LexUriServ/site/en/com/2005/com2005_0605en01.pdf

[65] European Commission, DG Health and Consumer Protection, Public Health, Health security and preparedness,
http://ec.europa.eu/health/ph_threats/com/Influenza/influenza_level_en.htm

[66] European Commission, DG Health and Consumer Protection, Public Health, *Influenza pandemic preparedness planning at EU level*,
http://ec.europa.eu/health/ph_threats/com/Influenza/influenza_level_en.htm

[67] European Centre for Disease Control and Prevention (ECDC), *Influenza — National Pandemic Influenza Plans*,
http://ecdc.europa.eu/Influenza/National_Influenza_Pandemic_Plans.php#EU_Countries

[68] World Health Organisation (WHO,) 'Avian influenza',
http://www.who.int/csr/disease/avian_influenza/en/index.html

[69] Berufsgenossenschaft Nahrungsmittel und Gaststätten, ‚Informationen zur Vogelgrippe — Besondere Hinweise für bestimmte Betriebsarten/Tätigkeiten',
http://praevention.portal.bgn.de/webcom/show_article.php/_c-7632/_nr-15/_p-2/i.html?PHPSESSID=d6d0cfd861f732cf3407e43672e7f226, 23rd May 2006

[70] Ministère des solidarities de la santé et de la famille, Direction générale de la santé, Bureau des maladies infectieuses et de la politique vaccinale: 'Conduite à tenir devant un foyer d'influenza aviaire à virus hautement pathogène et à risque établi de transmission humaine lors d'une épizootie en France ou dans les régions limitrophes', 15th Feb 2005, http://www.sante.gouv.fr/htm/dossiers/grippe_aviaire/protocole.pdf

[71] Ministère de l'agriculture et de la pêche, Ministère de l'emploi, de la cohésion sociale et du logement, Ministère des transports, de l'équipement, du tourisme et de la mer, 'Prévenir les risques liés à l'influenza aviaire', Jan 2006 http://www.grippeaviaire.gouv.fr/IMG/pdf/IA.pdf

[72] Dreller, S., Jatzwauk, J., Nassauer, A., Paszkiewicz, P., Tobys, H.-U., Rüden, H., 'Zur Frage des Atemschutzes vor luftübertragenen Infektionserregern (Investigations on suitable respiratory protection against airborne pathogens)', *Gefahrstoffe — Reinhaltung der Luft,* Vol. 66, No. 1-2, 2006, http://www.hvbg.de/d/bia/pub/grl/2006_003.pdf

[73] Booth, T., Kournikakis, B., Bastien, N., Ho, J., et al., 'Detection of airborne Severe Acute Respiratory Syndrome (SARS) Coronarovirus and environmental contamination in SARS 0utbreak units'. *JID* Vol. 191, No. 9, 2005, pp. 1472–7.

[74] Benbrook, C., Benbrook Consultant Services, USA, Sept 2004, http://depts.washington.edu/pnash/conf04/4_Presentations/13_Benbrook_Worker_Health_Standards.pdf

[75] Formenty, P., Roth, C., Gonzalez-Martin, F., Grein, T., et al., 'Les pathogènes émergents, la veille internationale et le nouveau règlement sanitaire international'. *Med Mal Infect* Vol. 35, No. 11, 2005, pp. 1–4.

[76] European Working Group for Legionella Infections — EWGLI, http://www.ewgli.org/

[77] European Network for Diagnostics of 'Imported' Viral Diseases — ENIVD, http://www.enivd.de/

[78] National Institute of Allergy and Infectious Diseases (NIAID), USA, 'The problem of antibiotic resistance', April 2004, http://www.niaid.nih.gov

[79] World Health Organization (WHO), 'Antimicrobial resistance', Fact sheet No. 194, revised January 2002.

[80] Bavdekar, S. B., 'Antibiotic resistance: unless we act soon!' *Journal of Postgraduate Medicine* Vol. 49, No. 2, pp. 107-18, April–June 2003, http://www.jpgmonline.com

[81] Livermore, D., Pillay, D., Cane, P., *Antimicrobial resistance — Inevitable but not unmanageable*. Health Protection Agency, 2005, http://www.hpa.org.uk/hpa/publications/amr_report_05/1_intro.htm

[82] Canadian Centre for Occupational Health and Safety. OSH Answers, 'Protecting health workers from Drug Resistant Organisms'. Health and Safety report. Vol. 1, No. 4, April 2003, http://www.ccohs.ca/newsletters/hsreport/issues/2003/04/ezine.html#oshanswers

[83] World Health Organisation, Alliance for the prudent use of antibiotics, 'Antibiotic resistance: synthesis of recommendations by expert policy groups', 2001, http://www.who.int

[84] Health Canada, 'Antibiotic resistance. It's Your Health', October 2005, http://www.hc-sc.gc.ca/

[85] Canadian Centre for Occupational Health and Safety, OSH Answers, 'Drug-resistant organisms?', March 2003, http://www.ccohs.ca/oshanswers/biol_hazards/drugresist.html

[86] American Pharmaceutical Association, USA, 'Combating antibiotic resistance — A continuing education program for pharmacists', 2001, http://www.carleton.ca/biology/biol1004/pdf/americanpharma2001.pdf

[87] De gezondheidssite voor Vlaanderen: 'Dossier antibiotica', Belgium, www.gezondheid.be, 2000–2005

[88] Wallinga, D., Bermudez, N., Hopkins, E., *Poultry on antibiotics: hazards to human health*. Institute for Agriculture and Trade Policy, USA, December 2002, http://www.mindfully.org/Food/Poultry-Antibiotics-HealthDec02.htm

[89] The campaign to end antibiotic overuse, USA. 'Keep antibiotics working: myths and realities', 2004, www.keepantibioticsworking.com

[90] Wallinga, D., 'Antimicrobial use in animal feed: an ecological and public health problem', *Minnesota Medicine* Vol. 85, October 2002 (Minnesota Medical Association), http://www.mmaonline.net/publications/MNMed2002/October/Wallinga.html

[91] 'Ban on antibiotics as growth promoters in animal feed enter into effect', European Commission, Belgium, 22nd December 2005, press release, IP/05/1687, http://europa.eu.int/rapid/pressReleasesAction.do?reference=IP/05/1687&format=HTML&aged=0&language=EN&guiLanguage=en

[92] Regulation (EC) No 1831/2003 of the European Parliament and of the Council of 22nd September 2003 on additives for use in animal nutrition, *Official Journal* L 268, 18/10/2003 pp. 29–43, http://europa.eu.int/eur-lex/lex/LexUriServ/LexUriServ.do?uri=CELEX:32003R1831:EN:HTML

[93] Chapin, A., Rule, A., Gibson, K., Buckley, T., Schwab K., 'Airborne Multidrug-Resistant Bacteria Isolated from a Concentrated Swine Feeding Operation', *Environmental Health Perspectives* Vol. 113, No. 2, February 2005, http://www.ehponline.org/members/2004/7473/7473.html

[94] Bryskier, A., 'Viridans group streptococci: a reservoir of resistant bacteria in oral cavities', *Clin Microbiol Infect* Vol. 8, 2002, pp. 65–9.

[95] World Health Organisation: 'WHO Global Task Force outlines measures to combat XDR-TB worldwide', 17th October 2006, http://www.who.int/mediacentre/news/notes/2006/np29/en/index.html

[96] European Centre for Disease Prevention and Control, Manissero D., Fernandez de la Hoz K., 'Surveillance methods and case definition for extensively drug resistant TB (XDR-TB) and relevance to Europe: summary update', *Eurosurveillance* Vol. 11, No. 11, 3rd November 2006, http://www.eurosurveillance.org/ew/2006/061103.asp#1

[97] Lõivukene, K., Kermes, K., Sepp, E., Adamson, V., Mitt, P., Kallandi, Ü., Otter, K., Naaber, P., 'Surveillance of antimicrobial resistance of invasive pathogens: the Estonian experience', *Eurosurveillance* Vol. 11, No. 2, February 2006, http://www.eurosurveillance.org/em/v11n02/1102-225.asp

[98] *Faisabilité de l'évaluation, chez les travailleurs d'abattoirs, des risques reliés à l'utilisation industrielle des antibiotiques dans l'élevage des animaux de consommation.* Institut de recherche en santé et en sécurité du travail du Québec (IRSST), Canada, 1984, http://www.irsst.qc.ca/files/documents/PubIRSST/PR-019.pdf

[99] Benbrook, C., *Developing worker health standards in sustainable agriculture*. Benbrook Consultant Services, USA, September 2004,

http://depts.washington.edu/pnash/conf04/4_Presentations/13_Benbrook_Worker_Health_Standards.pdf

[100] Sibergeld, E. K., *Perspectives on the science: what do we know and what do we need to know about microbiological risks of food animal production?* Johns Hopkins Bloomberg School of Public Health, USA, December 2004

[101] Canadian Centre for Occupational Health and Safety. OSH Answers: *Methicillin-Resistant Staphylococcus Aureus*, September 2005,
http://www.ccohs.ca/oshanswers/biol_hazards/methicillin.html

[102] van den Bogaard, A. E., London, N., Driessen, C., Stobberingh E. E., 'Antibiotic resistance of faecal Escherichia coli in poultry, poultry farmers and poultry slaughterers', *Journal of Antimicrobial Chemotherapy*, The Netherlands, Vol. 47, 2001, pp. 763–71, http://jac.oxfordjournals.org/cgi/content/full/47/6/763

[103] American Public Health Association, Washington, DC; Center for Science in the Public Interest, Washington, DC; Environmental Defense, Washington, DC; Global Resource Action Center for the Environment, New York, NY; Institute for Agriculture and Trade Policy, Minneapolis, MN; Natural Resources Defense Council, Washington, DC; Science and Environmental Health Network, Boston, MA; Sierra Club, Washington, DC; Union of Concerned Scientists. Union of concerned scientists, 'Limit Vital Antibiotics in Factory Farm Effluent' — Letter to the EPA, August 2003,
http://www.ucsusa.org/food_and_environment/antibiotics_and_food/letter-to-epa-urging-limits-on-antibiotics.html

[104] Meade-Callahan, M. J., 'Microbes: What They Do & How Antibiotics Change Them', American Institute of Biological Sciences. *ActionBioscience*, January 2001,
http://www.actionbioscience.org/evolution/meade_callahan.html

[105] World Health Organisation, 'WHO global strategy for containment of antimicrobial resistance', 2001, WHO/CDS/CSR/DRS/2001.2.,
http://www.who.int/csr/resources/publications/drugresist/WHO_CDS_CSR_DRS_2001_2_EN/en/

[106] Health Care Without Harm, USA, 'Antibiotic resistance and agricultural overuse of antibiotics — What health care food systems can do'. 2nd July 2004, www.noharm.org

[107] Mackiewicz, B., 'Study on exposure of pig farm workers to bioaerosols, immunologic reactivity and health effects', *Ann Agric Environ Med* Vol. 5, 1998, pp. 169–75.

[108] European Commission Directorate-General for Research, *Antibiotics resistance — A growing threat.* http://ec.europa.eu/research/leaflets/antibiotics/index_en.html

[109] National Center for Infectious Diseases, USA, *An ounce of prevention: keeps the germs away — Wash your hands often*, 2000,
http://www.cdc.gov/ncidod/op/_resources/OOP%20Brochure%2012.20.05.pdf

[110] Better Health, USA, *Antibiotic resistant bacteria*, 28th May 2001,
http://www.betterhealth.vic.gov.au

[111] American College of Physicians, USA, *What you can do to reduce the threat of antibiotic resistance.* http://www.acponline.org

[112] Canada's National Occupational Health & Safety Resource, Canada, *Drug-resistant organisms*, http://www.ccohs.ca, 2003

[113] Nicolle, L. E., *Infection control programmes to contain antimicrobial resistance*. World Health Organisation, Department of Communicable Disease Surveillance and Response, 2001, http://www.who.int

[114] Boyce, J. M., Pittet, D., 'Guideline for hand hygiene in health-care settings: recommendations of the healthcare infection control practices advisory committee and the HICPAC/SHEA/APIC/IDSA hand hygiene task force', *Infection control and hospital epidemiology*, USA, Vol. 23, No. 12, suppl. S3–S41, May 2003, http://www.ncbi.nlm.nih.gov/entrez/query.fcgi?cmd=Retrieve&db=PubMed&list_uids=12418624&dopt=Abstract

[115] Saiman, L., Siegel, J., 'Infection control and hospital epidemiology: Infection control recommendations for patients with cystic fibrosis: microbiology, important pathogens, and infection control practices to prevent patient-to-patient transmission', Cystic Fibrosis Foundation Consensus Conference on Infection Control Participants, Vol. 24 No. 5, S6–S53, USA, May 2003

[116] Muto, C. A., Jernigan, J. A., Ostrowsky, B. E., Richet, H. M., Jarvis, W. R., Boyce, J. M., Farr, B. M., 'SHEA guideline for preventing nosocomial transmission of multidrug-resistant strains of staphylococcus aureus and enterococcus', *Infection control and hospital epidemiology*, USA, Vol. 24, No. 5, May 2003, pp. 362–86.

[117] Zoutman, D. E., Ford, B. D., *The relationship between hospital infection surveillance and control activities and antibiotic-resistant pathogen rates*. Canadian Hospital Epidemiology Committee, Canadian Nosocomial Infection Surveillance Program, Health Canada, 2005, http://www.hc-sc.gc.ca/

[118] Centers for Disease Control and Prevention, USA, 'Hand hygiene guidelines fact sheet', October 2002, http://www.cdc.gov/

[119] Centers for Disease Control and Prevention, USA, 'Information about MRSA for healthcare personnel', August 2004, http://www.cdc.gov/ncidod/dhqp/ar_mrsa_healthcareFS.html

[120] Sulsky, S. I., Birk, T., Cohen, L. C., Luippold, R. S., Heidenreich, M. J., Nunes, A., *Effectiveness of measures to prevent needlestick injuries among employees in health professions*. Hauptverband der gewerblichen Berufsgenossenschaften (HVBG), Sankt Augustin, 2006, www.hvbg.de Webcode: 1961356

[121] World Health Organisation, Department of Communicable Disease Surveillance and Response, *Prevention of hospital-acquired infections — A practical guide* (2nd edn), 2002, http://www.who.int

[122] Evans, M. C., Wegener, H. C., 'Antimicrobial Growth Promoters and Salmonella spp., Campylobacter spp. in Poultry and Swine. Centre for Disease Control and Prevention', *Emerging Infectious Diseases*, Denmark, Vol. 9, No. 4, April 2003, http://www.cdc.gov/ncidod/EID/vol9no4/pdfs/02-0325.pdf

[123] 'Impacts of antimicrobial growth promoter termination in Denmark', The WHO international review panel's evaluation of the termination of the use of antimicrobial growth promoters in Denmark. World Health Organisation, Department of Communicable Diseases, Prevention and Eradication, Collaborating Centre for Antimicrobial Resistance in Foodborne Pathogens, Denmark, November 2002, WHO/CDS/CPE/ZFK/2003.1. http://whqlibdoc.who.int/hq/2003/WHO_CDS_CPE_ZFK_2003.1.pdf

[124] 'Danish ban on antibiotics proves successful.' *Food Production Daily*, 5th May 2003, http://www.foodproductiondaily.com/news/ng.asp?id=29771-danish-ban-on

[125] Todar, K., 'Mechanisms of bacterial pathogenicity: Endotoxins'. Todar's online textbook of bacteriology. http://textbookofbacteriology.net/endotoxin.html, 2002

[126] Rylander, R., 'Organic dusts and disease: A continuous research challenge', *Am J Ind Med* Vol. 46, No. 4, 2004, pp. 323–6.

[127] Eduard, W., Westby, M. H., Larsson, L., 'Solubility of endotoxins from Escherichia coli and Pseudomonas aeruginosa', *Am J Ind Med* Vol. 46, No. 4, 2004, pp. 375–7.

[128] Radon, K., 'The two sides of the endotoxin coin', *Occup Environ Med* Vol. 63, No. 1, 2006, pp. 73–8.

[129] Spaan, S., Wouters, I. M., Oosting, I., Doekes, G., Heederik, D., 'Exposure to inhalable dust and endotoxins in agricultural industries', *J Environ Monit* Vol. 8, No. 1, 2006, pp. 63–72.

[130] Omland, O., 'Exposure and respiratory health in farming in temperate zones — A review of the literature', *Ann Agric Environ Med* Vol. 9, No. 2, 2002, pp. 119–36.

[131] Malmberg, P., Rask-Andersen, A., Hoglund, S., Kolmodin-Hedman, B., Read, G. J., 'Incidence of organic dust toxic syndrome and allergic alveolitis in Swedish farmers', *Int Arch Allergy Appl Immunol* Vol. 87, No. 1, 1988, pp. 47–54.

[132] Monso, E., Magarolas, R., Radon, K., Danuser, B., Iversen, M., Weber, C., et al., 'Respiratory symptoms of obstructive lung disease in European crop farmers', *Am J Respir Crit Care Med* Vol. 162(4 Pt 1), 2000, pp. 1246–50.

[133] Melbostad, E., Eduard, W., 'Organic dust-related respiratory and eye irritation in Norwegian farmers', *Am J Ind Med* Vol. 39, No. 2, 2001, pp. 209–17.

[134] Linaker, C., Smedley, J., 'Respiratory illness in agricultural workers', *Occup Med* Vol. 52, No. 8, 2002, pp. 451–9.

[135] Gora, A., Skorska, C., Prazmo, Z., Krysinska-Traczyk, E., Sitkowska, J., Dutkiewicz, J., 'Exposure to bioaerosols: Allergic reactions and respiratory function in Polish hop growers', *Am J Ind Med* Vol. 46, No. 4, 2004, pp. 371–4.

[136] Monso, E., Riu, E., Radon, K., Magarolas, R., Danuser, B., Iversen, M., et al., 'Chronic obstructive pulmonary disease in never-smoking animal farmers working inside confinement buildings', *Am J Ind Med* Vol. 46, No. 4, 2004, pp. 357–62.

[137] Zejda, J. E., Barber, E., Dosman, J. A., Olenchock, S. A., McDuffie, H. H., Rhodes, C., et al., 'Respiratory health status in swine producers relates to endotoxin exposure in the presence of low dust levels', *J Occup Med* Vol. 36, No. 1, 1994, pp. 49–56.

[138] Rylander, R., Carvalheiro, M. F., 'Airways inflammation among workers in poultry houses', *Int Arch Occup Environ Health* Jan 2006, pp. 1–4.

[139] Harper, M., Andrew, M.E., 'Airborne endotoxin in woodworking (joinery) shops', *J Environ Monit* Vol. 8, No. 1, Jan 2006, pp. 73–8.

[140] Chattopadhyay, B. P., Saiyed, H. N., Mukherjee, A. K., 'Byssinosis among jute mill workers', *Ind Health* Vol. 41, No. 3, 2003, pp. 265–72.

[141] Mukherjee, A. K., Chattopadhyay, B. P., Bhattacharya, S. K., Saiyed, H. N., 'Airborne endotoxin and its relationship to pulmonary function among workers in an Indian jute mill', *Archives of Environmental Health* Vol. 59, No. 4, 2005, pp. 202–8.

[142] Hang, J., Zhou, W., Wang, X., Zhang, H., Sun, B., Dai, H., et al., 'Microsomal epoxide hydrolase, endotoxin, and lung function decline in cotton textile workers', *Am J Respir Crit Care Med* Vol. 171, No. 2, 2005, pp. 165–70.

[143] Torres Costa, J., Ferreira, J. A., Castro, E., Vaz, M., Barros, H., Marques, J. A., 'One-week variation of cotton dust and endotoxin levels in a cotton mill. Relation with the daily variation of the expiratory flow rates', *Acta Med Port* Vol. 17, No. 2, 2004, pp. 149–56.

[144] Wang, X. R., Zhang, H. X., Sun, B. X., Dai, H. L., Hang, J. Q., Eisen, E. A., et al., 'A 20-year follow-up study on chronic respiratory effects of exposure to cotton dust', *Eur Respir J* Vol. 26, No. 5, Nov 2005, pp. 881–6.

[145] Oldenburg, M., Latza, U., Baur, X., 'Airborne exposure parameters in a cotton spinning mill: Relevance and health effects', *Gefahrstoffe Reinhalt Luft* Vol. 63, No. 9, 2003, pp. 373–80.

[146] Oldenburg, M., Baur, X., 'Impairment of lung function as a result of exposure to endotoxins in a German cotton mill?' *Arbeitsmed Sozialmed Umweltmed* Vol. 38, No. 7, 2003, pp. 370–74.

[147] Su, H. J. J., Chen, H. L., Huang, C. F., Lin, C. Y., Li, F. C., Milton, D. K., 'Airborne fungi and endotoxin concentrations in different areas within textile plants in Taiwan: A 3-year study', *Environ Res* Vol. 89, No. 1, 2002, pp. 58–65.

[148] Nordness, M. E., Zacharisen, M. C., Schlueter, D. P., Fink, J. N., 'Occupational lung disease related to Cytophaga endotoxin exposure in a nylon plant', *J Occup Environ Med* Vol. 45, No. 4, 2003, pp. 385–92.

[149] Thorn, J., Kerekes, E., 'Health effects among employees in sewage treatment plants: A literature survey', *Am J Ind Med* Vol. 40, No. 2, 2001, pp. 170–79.

[150] Thorn, J., Beijer, L., 'Work-related symptoms and inflammation among sewage plant operatives', *Int J Occup Environ Health* Vol. 10, No. 1, 2004, pp. 84–9.

[151] Smit, L. A. M., Spaan, S., Heederik, D., 'Endotoxin exposure and symptoms in wastewater treatment workers', *Am J Ind Med* Vol. 48, No. 1, 2005, pp. 30–39.

[152] Heldal, K. K., Eduard, W., 'Associations between acute symptoms and bioaerosol exposure during the collection of household waste', *Am J Ind Med* Vol. 46, No. 3, 2004, pp. 253–60.

[153] Gladding, T., Thorn, J., Stott, D., 'Organic dust exposure and work-related effects among recycling workers', *Am J Ind Med* Vol. 43, No. 6, 2003, pp. 584–91.

[154] Kennedy, S. M., Copes, R., Bartlett, K. H., Brauer, M., 'Point-of-sale glass bottle recycling: Indoor airborne exposures and symptoms among employees', *Occup Environ Med* Vol. 61, No. 7, 2004, pp. 628–35.

[155] Sigsgaard, T., Jensen, L. D., Abell, A., Wurtz, H., Thomsen, G., 'Endotoxins isolated from the air of a Danish paper mill and the relation to change in lung function: An 11-year follow-up', *Am J Ind Med* Vol. 46, No. 4, 2004, pp. 327–32.

[156] Kraus, T., Pfahlberg, A., Zobelein, P., Gefeller, O., Raithel, H. J., 'Lung function among workers in the soft tissue paper-producing industry', *Chest* Vol. 125, No. 2, 2004, pp. 731–6.

[157] Prazmo, Z., Dutkiewicz, J., Skorska, C., Sitkowska, J., Cholewa, G., 'Exposure to airborne gram-negative bacteria, dust and endotoxin in paper factories', *Ann Agric Environ Med* Vol. 10, No. 1, 2003, pp. 93–100.

[158] Korhonen, K., Liukkonen, T., Ahrens, W., Astrakianakis, G., Boffetta, P., Burdorf, A., et al., 'Occupational exposure to chemical agents in the paper industry', *Int Arch Occup Environ Health* Vol. 77, No. 7, 2004, pp. 451–60.

[159] Gorny, R. L., Szponar, B., Larsson, L., Pehrson, C., Prazmo, Z., Dutkiewicz, J., 'Metalworking fluid bioaerosols at selected workplaces in a steelworks', *Am J Ind Med* Vol. 46, No. 4, 2004, pp. 400–403.

[160] Taibjee, S. M., Foulds, I. S., 'Microorganism-induced skin disease in workers exposed to metalworking fluids [1]', *Occup Med* Vol. 53, No. 7, 2003, pp. 483–4.

[161] Linnainmaa, M., Kiviranta, H., Laitinen, J., Laitinen, S., 'Control of workers' exposure to airborne endotoxins and formaldehyde during the use of metalworking fluids', *Am Ind Hyg Assoc J* Vol. 64, No. 4, 2003, pp. 496–500.

[162] Park, D., Choi, B., Kim, S., Kwag, H., Joo, K., Jeong, J., 'Exposure assessment to suggest the cause of sinusitis developed in grinding operations utilizing soluble metalworking fluids', *Journal of Occupational Health* Vol. 47, No. 4, 2005, pp. 319–26.

[163] Bang, B., Aasmoe, L., Aamodt, B. H., Aardal, L., Andorsen, G. S., Bolle, R., et al., 'Exposure and airway effects of seafood industry workers in northern Norway', *Journal of Occupational and Environmental Medicine* Vol. 47, No. 5, 2005, pp. 482–92.

[164] Brasche, S., Bullinger, M., Petrovitch, A., Mayer, E., Gebhardt, H., Herzog, V., et al., 'Self-reported eye symptoms and related diagnostic findings — comparison of risk factor profiles', *Indoor Air* Vol. 15, 2005, pp. 56–64.

[165] Platts-Mills, J., Custis, N., Kenney, A., Tsay, A., Chapman, M., Feldman S., et al., 'The effects of cage design on airborne allergens and endotoxin in animal rooms: high-volume measurements with an ion-charging device', *Contemp Top Lab Anim Sci* Vol. 44, No. 2, 2005, pp. 12–16.

[166] Ding, J. L., Ho, B., 'A new era in pyrogen testing', *Trends Biotechnol.* Vol. 19, No. 8, 2001, pp. 277–81.

[167] Vercelli, D., 'Genetics, epigenetics, and the environment: switching, buffering, releasing', *J Allergy Clin Immunol* Vol. 113, No. 3, 2004, pp. 381–6, http://www.ncbi.nlm.nih.gov/entrez/query.fcgi?cmd=Retrieve&db=PubMed&list_uids=15007332&dopt=Abstract

[168] Vercelli, D., 'Genetic polymorphism in allergy and asthma', *Curr Opin Immunol* Vol. 15, No. 6, Dec 2003, pp. 609–13, http://www.ncbi.nlm.nih.gov/entrez/query.fcgi?cmd=Retrieve&db=PubMed&list_uids=14630192&dopt=Abstract

[169] BMWA, 'Irritativ-toxische Wirkungen von luftgetragenen biologischen Arbeitsstoffen am Beispiel der Endotoxine', BArbBl., 6, 45–59, Mai 2005, *Informationspapier aus Bekanntmachung des BMWA*, III B 3-34504 — 7.

[170] Amoureux, M. C., 'Pathophysiological role of endotoxins, a common denominator to various diseases', *Pathol Biol* Paris, Vol. 52, No. 7, Sep 2004, pp. 415–22.

[171] Yang, C.S., *Endotoxins*. http://www.stl-inc.com/technical_Papers/endotoxin.pdf, 2004

[172] Natural Resources Defense Council, 'Health effects of endotoxin', 2006, http://www.nrdc.org/health/effects/katrinadata/endotoxin.asp

[173] Danuser, B., Monn, C., 'Endotoxins in the workplace and in the environment', *Schweiz Med Wochenschr* Vol. 129, No. 12, March 1999, pp. 475–83.

[174] Radon, K., Ehrenstein, V., Praml, G., Nowak, D., 'Childhood visits to animal buildings and atopic diseases in adulthood: An age-dependent relationship', *Am J Ind Med* Vol. 46, No. 4, 2004, pp. 349–56.

[175] Liu, A. H., Redmon, J., 'Endotoxin: friend or foe?' *Allergy Asthma Proc* Vol. 22, No. 6, 2001, pp. 337–40.

[176] Vandenbulcke, L., Bachert, C., Van, C. P., Claeys, S., 'The innate immune system and its role in allergic disorders', *Int Arch Allergy Immunol* Vol. 139, No. 2, 2006, pp. 159–65.

[177] Eduard, W., Omenaas, E., Bakke, P. S., Douwes, J., Heederik, D., 'Atopic and non-atopic asthma in a farming and a general population', *Am J Ind Med* Vol. 46, No. 4, 2004, pp. 396–9.

[178] Heederik, D., Sigsgaard, T., 'Respiratory allergy in agricultural workers: recent developments', *Curr Opin Allergy Clin Immunol* Vol. 5, No. 2, 2005, pp. 129–34.

[179] Song, B. J., Liu, A. H., 'Metropolitan endotoxin exposure, allergy and asthma', *Curr Opin Allergy Clin Immunol* Vol. 3, No. 5, Oct 2003, pp. 331–5.

[180] Smit, L. A., Wouters, I. M., Hobo, M. M., Eduard, W., Doekes, G., Heederik, D., 'Agricultural seed dust as a potential cause of organic dust toxic syndrome', *Occup Environ Med* Vol. 63, No. 1, Jan 2006, pp. 59-67

[181] Eduard, W., Douwes, J., Omenaas, E., Heederik, D., 'Do farming exposures cause or prevent asthma? Results from a study of adult Norwegian farmers', *Thorax* Vol. 59, No. 5, May 2004, pp. 381–6.

[182] Laitinen, S., Kangas, J., Husman, K., Susitaival, P., 'Evaluation of exposure to airborne bacterial endotoxins and peptidoglycans in selected work environments', *Ann Agric Environ Med* Vol. 8, No. 2, 2001, pp. 213–9.

[183] Castor, M. L., Wagstrom, E. A., Danila, R. N., Smith, K. E., Naimi, T. S., Besser, J. M., et al., 'An outbreak of Pontiac fever with respiratory distress among workers performing high-pressure cleaning at a sugar-beet processing plant', *Journal of Infectious Diseases* Vol. 191, No. 9, 2005, pp. 1530-37.

[184] Hend, I. M., Milnera, M., Milnera, S. M., 'Bactericidal treatment of raw cotton as the method of byssinosis prevention', *Am Ind Hyg Assoc J* Vol. 64, No. 1, 2003, pp. 88–94.

[185] Lange, J. H., Fedeli, U., Mastrangelo, G., 'Is mushroom workers' chronic cough the same as byssinosis and what should the occupational exposure limit be for endotoxin?' *Chest* Vol. 123, No. 6, 2003, pp. 2160–61.

[186] Wang, X. R., Eisen, E. A., Zhang, H. X., Sun, B. X., Dai, H. L., Pan, L. D., et al., 'Respiratory symptoms and cotton dust exposure; results of a 15-year follow up observation', *Occup Environ Med* Vol. 60, No. 12, 2003, pp. 935–41.

[187] Herr, C. E., zur Nieden, A., Stilianakis, N. I., Gieler, U., Eikmann, T. F., 'Health effects associated with indoor storage of organic waste', *Int Arch Occup Environ Health* Vol. 77, No. 2, 2004, pp. 90–96.

[188] Poulsen, O. M., Breum, N. O., Ebbehoj, N., Hansen, A. M., Ivens, U. I., Van, L. D., et al., 'Collection of domestic waste. Review of occupational health problems and their possible causes', *Sci Total Environ* Vol. 170(1–2), Aug 1995, pp. 1–19.

[189] Heldal, K. K., Halstensen, A. S., Thorn, J., Diupesland, P., Wouters, I., Eduard, W., et al., 'Upper airway inflammation in waste handlers exposed to bioaerosols', *Occup Environ Med* Vol. 60, No. 6, 2003, pp. 444–50.

[190] Steiner, D., Jeggli, S., Tschopp, A., Bernard, A., Oppliger, A., Hilfiker, S., et al., 'Clara cell protein and surfactant protein B in garbage collectors and in wastewater workers exposed to bioaerosols', *Int Arch Occup Environ Health* Vol. 78, No. 3, 2005, pp. 189–97.

[191] Thorne, P. S., Duchaine, C., Douwes, J., Eduard, W., Gorny, R., Jacobs, R., Reponen, T., Schierl, R., Szponar, B., Working Group Report 4: 'Exposure assessment for biological agents', *Am J Ind Med* Vol. 46, No. 4, 2004, pp. 419–22, http://www.ncbi.nlm.nih.gov/entrez/query.fcgi?cmd=Retrieve&db=PubMed&list_uids=15376220&dopt=Abstract, 46:419-422

[192] Jacobs, R. R., Chun, D., 'Inter-laboratory analysis of endotoxin in cotton dust samples', *Am J Ind Med* Vol. 46, No. 4, 2004, pp. 333–7.

[193] Douwes, J., Thorne, P., Pearce, N., Heederik, D., 'Bioaerosol health effects and exposure assessment: progress and prospects', *Ann Occup Hyg* Vol. 47, No. 3, Apr 2003, pp. 187–200, http://annhyg.oxfordjournals.org/cgi/content/full/47/3/187, 47(3): 187–200

[194] Dutkiewicz, J., Gorny, R. L., 'Biologic factors hazardous to health: classification and criteria of exposure assessment', *Medycyna Pracy* Vol. 53, No. 1, 2002, pp. 29–39, http://www.ncbi.nlm.nih.gov/entrez/query.fcgi?cmd=Retrieve&db=PubMed&list_uids=12051150&dopt=Abstract

[195] Lane, S. R., Nicholls, P. J., Sewell R. D. E., 'The measurement and health impact of endotoxin contamination in organic dusts from multiple sources: focus on the cotton industry', *Inhal Toxicol* Vol. 16, No. 4, 2004, pp. 217–29.

[196] Liebers, V., Bruning, T., Raulf-Heimsoth, M., 'Occupational exposure to endotoxins and possible health effects', *Gefahrstoffe Reinhalt Luft* Vol. 64, No. 5, 2004, pp. 207–16.

[197] Heederik, D., Douwes, J., 'Towards an occupational exposure limit for endotoxins?' *Ann Agric Environ Med* Vol. 4, 1997, pp. 17–19, http://www.aaem.pl/pdf/9741_17.pdf

[198] Suva — Caisse nationale suisse d'assurance en cas d'accidents, *Valeurs limites d'exposition aux postes de travail 2005*. Référence: 1903.f. February 2005. http://wwwitsp1.suva.ch/sap/its/mimes/waswo/99/pdf/01903-f.pdf#search=%22SUVA%2C%20Valeurs%20limites%20d%E2%80%99exposition%20aux%20postes%20de%20travail%202005%22

[199] Rieger, M. A., Lohmeyer, M., Nubling, M., Neuhaus, S., Diefenbach, H., Hofmann, F., 'A description of the standardized measurement procedures and recommended threshold limit values for biological hazards in Germany', *J Agric Saf Health* Vol. 11, No. 2, 2005, pp. 185–91, http://asae.frymulti.com/abstract.asp?aid=18185&t=2

[200] Hawkins, L. D., Christ, W. J., Rossignol, D. P., 'Inhibition of endotoxin response by synthetic TLR4 antagonists', *Curr Top Med Chem* Vol. 4, No. 11, 2004, pp. 1147–71, http://www.ncbi.nlm.nih.gov/entrez/query.fcgi?cmd=Retrieve&db=PubMed&list_uids=15279606&dopt=Abstract

[201] Storm, J. F., Genter, M. B., *Respiratory Risks in Agriculture*. AG-MED-6, 1996, http://www.ces.ncsu.edu/depts/fcs/health/pubs/agmed6.pdf

[202] Palmberg, L., Larsson, B. M., Sundblad, B. M., Larsson, K., 'Partial protection by respirators on airways responses following exposure in a swine house', *Am J Ind Med* Vol. 46, No. 4, 2004, pp. 363–70, http://www.ncbi.nlm.nih.gov/entrez/query.fcgi?cmd=Retrieve&db=PubMed&list_uids=15376212&dopt=Abstract

[203] Paky, A., Knoblauch, A., 'Dust-exposure, dust-induced lung-disease and respiratory protective systems in farming', *Schweizerische Medizinische Wochenschrift* Vol. 125, No. 10, Mar 1995, pp. 458–66.

[204] Quezada, N. V., Lange, J. H. ' Final clearance criteria alter mould remediation', *Indoor Built Environ* Vol. 13, 2004, pp. 199–203, http://ibe.sagepub.com/cgi/content/abstract/13/3/199

[205] Gots, R. E., Layton, N. J., Pirages, S. W., 'Indoor health: background levels of fungi', *AIHA Journal*, http://www.ncbi.nlm.nih.gov/entrez/query.fcgi?cmd

[206] Deininger, C. et al., Biologische Einwirkungen. In: von Hahn, N., Kleine, H., 'Innenraumarbeitsplätze — Vorgehensempfehlungen für die Ermittlungen zum Arbeitsumfeld', *HVBG, 2. vollständig überarbeitete Auflage*, 248–270, Juli 2005

[207] Singh, J., 'Occupational Exposure to Moulds in Buildings', *Indoor Built Environ* Vol. 10, 2001, pp. 172–8, http://www.ebssurvey.co.uk/docs/MonitoringIndoorAirPollution.pdf

[208] Aydogdu, H., Asan, A., Otkun, M. T., Ture, M., 'Monitoring of Fungi and bacteria in the Indoor Air of Primary Schools in Edirne City, Turkey', *Indoor Built Environ* Vol. 14, No. 5, 2005, pp. 411–25.

[209] Husman, T., 'Health Effect of Indoor-Air Microorganisms', *Scand J Work Environ Health* Vol. 22, 1996, pp. 5–13.

[210] Lugauskas, A., Krikstaponis, A., 'Filamentous Fungi Isolated in hospitals and Some Medical Institutions in Lithuania', *Indoor Built Environ* Vol. 13, 2004, pp. 101–8.

[211] Blome, H. et al., 'Method for measurement of the mould fungus concentration in workplace atmospheres', Kennzahl 9420, 30th suppl. IV/03 In: Messung von Gefahrstoffen — BIA-Arbeitsmappe. Published by: Berufsgenossenschaftliches Institut für Arbeitsschutz — BIA. Erich Schmidt Verlag, Berlin — loose-leaf edition, 2003

[212] Wu, Z., Blomquist, G., Westermark, S.-O., Wang, X.-R., 'Application of PCR and probe hybridization techniques in detection of airborne fungal spores in environmental samples', *J Environ Monit* Vol. 4, 2002, pp. 673–8.

[213] Wu, Z., Wang, X.-R., Blomquist, G., 'Evaluation of PCR-primers and PCR-Conditions for specific detection of common airborne fungi', *J Environ Monit* Vol. 4, 2004, pp. 377–82.

[214] Wu, Z., Tsumura, Y., Blomquist, G., Wang, X.-R., '18S rRNA Gene Variation among Common Airborne Fungi and Development of Specific Oligonucleotide Probes for the Detection of Fungal Isolates', *Applied and Environmental Microbiology* Vol. 69, No. 9, 2003, pp. 5389–97.

[215] Zeng, Q.-Y., Wang, X.-R., Blomquist, G., 'Development of mitochondrial SSU rDNA-based oligonucleotide probes for specific detection of common airborne fungi', *Molecular and Cellular Probes* Vol. 17, 2003, pp. 281–8, http://www.bls.gov/oco/ocos256.htm

[216] Lange, J. H., 'Mould: The next Environmental Industry', *Indoor Built Environ* Vol. 13, 2004, pp. 165–7.

[217] 'Schimmelpilze in Innenräumen — Nachweis, Bewertung, Qualitätsmanagement', Arbeitskreis 'Qualitätssicherung — Schimmelpilze in Innenräumen'. Stuttgart, 2001

[218] Leitfaden zur Vorbeugung, Untersuchung, Bewertung und Sanierung von Schimmelpilzwachstum in Innenräumen ('Schimmelpilz-Leitfaden'). Umweltbundesamt, Innenraumlufthygienekommission, Berlin, 2002

[219] Kolk, A. et al., 'Biostoffliste — Handlungshilfen zur Gefährdungsbeurteilung beim Umgang mit biologischen Arbeitsstoffen', *HVBG*, Nov 2005

[220] J. Krause, Indoor Air Solutions Inc., *Dissecting the practice of using bioaerosol samples for evaluating mold remediation efficacy*. American Industrial Hygiene Association (AIHA), 2004, Paper 250 of Podium 131: Mold Remediation, http://www.aiha.org/abs04/po131.htm

[221] Bartlett, K. H., Kennedy, S. M., Brauer, M., Van Netten, C., Dill, B., 'Evaluation and a predictive model of airborne fungal concentrations in school classrooms', *Ann Occup Hyg* Vol. 48, No. 6, 2004, pp. 547–54, http://www.ncbi.nlm.nih.gov/entrez/query.fcgi?cmd=Retrieve&db=PubMed&list_uids=15302620&dopt=Abstract

[222] Fung, F., Hughson, W. G., 'Health effects of indoor fungal bioaerosol exposure', *Applied Occupational and Environmental Hygiene* Vol. 18, 2003, pp. 535–44, http://www.ncbi.nlm.nih.gov/entrez/query.fcgi?cmd=Retrieve&db=PubMed&list_uids=12791550&dopt=Abstract

[223] Canadian Construction Association, *Mould guidelines for the Canadian construction industry*, 2004, Standard Construction Document CCA 82. http://www.cca-acc.com/documents/electronic/cca82/cca82.pdf

[224] US Department of Labor, Occupational Safety and Health Administration, *Indoor air quality investigation*. OSHA Technical Manual (OTM), January 1999, chapter 2, http://www.osha.gov/dts/osta/otm/otm_toc.html

[225] Commission of the European Communities, *Biological particles in indoor environments*, Report 12, 1993, http://www.jrc.cec.eu.int/pce/eca_reports/ECA_Report12.pdf

[226] Robbins, C. A., Swenson, L. J., Nealley, M. L., Gots, R. E., Kelman, B. J., 'Health effect of mycotoxins in indoor air: A critical review', *Applied Occupational and Environmental Hygiene* Vol. 15, No. 10, 2000, pp. 773–84. http://taylorandfrancis.metapress.com/openurl.asp?genre=article&id=doi:10.1080/1047322005012941

[227] Kemp, P. C., Neumeister-Kemp, H.-g., Esposito, B., Lysek, G., Murray, F., 'Changes in airborne fungi from the outdoors to indoor air; large HVAC systems in nonproblem buildings in two different climates', *AIHA Journal* Vol. 64, 2003, pp. 269–75.

[228] Menetrez, M.Y., Foarde, K. K., 'Research and development of prevention and control measures for mold contamination', *Indoor Built Environ* Vol. 13, 2004, pp. 109–14, http://ibe.sagepub.com/cgi/content/abstract/13/2/109

[229] 'Adverse human health effects associated with molds in the indoor environment', *JOEM* Vol. 45, No. 5, 2003, pp. 470–78.

[230] Canadian Construction Association, *Mould — an informational brochure*, http://www.cca-acc.com/mould/brochure/ccamouldbrochure.pdf

[231] Bholah, R., Subratty, A.H., 'Indoor biological contaminants and symptoms of sick building syndrome in office buildings in Mauritius', *International Journal of Environment Health Research* Vol. 12, 2002, pp. 93–8.

[232] Sessa, R., Di Pietro, M., Schia Voni, G., Santino I., Atieri A., Pinelli, S., Del Piano, M., 'Microbiological Indoor Air Quality in Healthy Building', *Microbiologica* Vol. 25, 2002, pp. 51–6.

[233] Innenraumlufthygiene-Kommission des Umweltbundesamtes, '*Leitfaden für die Innenraumlufthygiene in Schulgebäuden*', Umweltbundesamt, Berlin, Juni 2000. http://www.umweltdaten.de/publikationen/fpdf-l/1824.pdf

[234] Kansas State University, '*Agricultural worker respiratory hazards*'. Occupational health — Fact sheet 39. http://www.k-state.edu/research/comply/iacuc/occhs/fact39.html

[235] Centers for Disease Control and Prevention (CDC), '*Mold: A growing problem*'. Testimony of Dr. Stephen Redd, Chief, Air and Respiratory Branch, accompanied by Tom Sinks both of CDC's National Center for Environmental Health, before the House Financial Services Housing and Community Opportunity and Oversight and Investigations Subcommittees. http://www.cdc.gov/washington/testimony/eh071802.htm, July 2002

[236] '*Hazardous material removal workers*'. US Department of Labor. Bureau of Labor Statistics, US Department of Labor. Occupational Outlook Handbook, 2006–07 Edition, Dec 2005, http://www.bls.gov/oco/ocos256.htm

[237] Singh, J., 'Toxic moulds and Indoor Air Quality', *Indoor Built Environ* Vol. 14, No. 3-4, 2005, pp. 229–34.

[238] Mussalo-Rauhamaa, H., Suomalainen, H., Helin, T., Elg, P., Orpana, A., 'Bronchoalveolar lavage findings from persons exposed to moulds in water-damaged houses', *Indoor Built Environ* Vol. 12, 2003, pp. 235–37.

[239] Matheson, M. C., Abramson, M. J, Dharmage, S. C., Forbes, A. B., Raven, J. M., Thien, F. C. K., Walters, E. H., 'Changes in indoor allergen and fungal levels predict changes in asthma activity among young adults', *Clin Exp Allergy* Vol. 35, 2005, pp. 907–13.

[240] Reijula, K., 'Moisture-problem buildings with molds causing work-related diseases', *Adv Appl Microbiol* Vol. 55, 2004, pp. 175–89.

[241] Miller, V., 'Understanding Mycotoxin Testing and Interpretation', *Occupational Hazards*, 2003, pp. 49–52.

[242] Ebbehøj N. E., Hansen M. Ø., Sigsgaard T., Larsen, L., 'Building-related symptoms and molds: a two-step intervention study', *Indoor Air* Vol. 12, No. 4, December 2002, p. 273, http://www.blackwell-synergy.com/doi/abs/10.1034/j.1600-0668.2002.02141.x

[243] 'Council Directive 1999/31/EC on the landfill of waste', *Official Journal* L 182, 16/07/1999 pp. 1–19.
http://eur-lex.europa.eu/LexUriServ/LexUriServ.do?uri=CELEX:31999L0031:EN:HTML, 26 April 1999

[244] Health and Safety Executives — HSE, *Health and safety in the waste management and recycling industries*, May 2006, http://www.hse.gov.uk/waste/

[245] Poulsen, O. M. et al., 'Sorting and recycling of domestic waste: review of occupational health problems and their possible causes', *Sci Total Environ* Vol. 168, 1995, pp. 33-56, http://www.ncbi.nlm.nih.gov/entrez/query.fcgi?cmd=Retrieve&db=PubMed&list_uids=7610383&dopt=Abstract

[246] Institut National de Recherche et de Sécurité (INRS,) *Déchets et risques professionnels*, Dossier. 26 Jan 2006, http://www.inrs.fr/

[247] Bomel Limited, *Mapping health and safety standards in the UK waste industry*. Health and Safety Executive — HSE. Research report 240,
http://www.hse.gov.uk/research/rrpdf/rr240.pdf, ISBN 0 7176 2865 5, 2004

[248] European Agency for Safety and Health at Work, *Biological agents* (ISSN 1681–2123), Facts 41, 2003, Belgium,
http://osha.eu.int/publications/factsheets/41/factsn41-en.pdf

[249] Schappler-Scheele, B., Schürmann, W., Hartung, J., Missel, T., Benning, C., Schröder, H., Weber, J., *Study of the health risks of employees in compost preparation plants*.

Bremerhaven, Wirtschaftsverlag NW Verlag für neue Wissenschaft GmbH, http://www.baua.de/nn_5846/en/Practical-experience/Publications/Institute-Report-Series/Research-reports/1999/Fb_20844.html__nnn=true, Schriftenreihe der Bundesanstalt für Arbeitsschutz und Arbeitsmedizin: Forschungsbericht, Fb 844, Edition 1, 1999.

[250] Clark, C. S., Bjornson, H. S., Schwartz-Fulton, J., Holland, J. W., Gartside, P. S., 'Biological health risks associated with the composting of wastewater treatment plant sludge', *Journal of the Water Pollution Control Federation* Vol. 56, 1984, pp. 1269–76.

[251] Marth, E., Reinthaler, F.F., Schaffler, K., Jelovcan, S., Haselbacher, S., Eibel, U., Kleinhappl, B., 'Occupational health risks to employees of waste treatment facilities', *Ann Agric Env Med* Vol. 4, 1997, pp. 143–7.

[252] Ivens, U., Breum, N. O., Ebbehoj, N., Nielsen, B. H., Poulsen, O., Wurtz, H., 'Exposure-response relationship between gastrointestinal problems among waste collectors and bioaerosol exposure', *Scand J Work Environ Health* Vol. 25, 1999, pp. 238–45.

[253] Rylander, R., 'Organic dusts — from knowledge to prevention', *Scand J Work Environ Health* 20 special issue, 1994, pp. 116–22.

[254] Dutkiewicz, J., 'Bacteria and fungi in organic dust as potential health hazard', *Ann Agric Environ Med* Vol. 4, 1997, pp. 11–16, http://www.aaem.pl/pdf/9741_11.pdf, 4, 11-16.

[255] Nielsen, E. M. et al., 'Bioaerosol exposure in waste collection: a comparative study on the significance of collection equipment, type of waste and seasonal variation', *Ann Occup Hyg* Vol. 41, No. 3, 1997, pp. 325–44; *J Work Environ Health* Vol. 25, No. 3, 1999, pp. 238–45.

[256] Nielsen, E. M., et al., 'Occupational bioaerosol exposure during collection of household waste', *Ann Agric Environ Med* Vol. 2, 1995, pp. 53–9.

[257] Kiviranta, H. et al., 'Exposure to airborne microorganisms and volatile organic compounds in different types of waste handling', *Ann Agric Environ Med* Vol. 6, 1999, pp. 39–44.

[258] Krajewski, J. A. et al., 'Occupational exposure to organic dust associated with municipal waste collection and management', *Int J Occup Med Environ Health* Vol. 15, No. 3, 2002, pp. 289–301.

[259] Yang, C.-Y. et al., 'Adverse health effects among household waste collectors in Taiwan', *Environ Res* Section a 85, 2001, pp. 195–99.

[260] Lavoie, J., Guertin, S., *Étude des agents biologiques et des contraintes ergonomiques lors de l'utilisation de camions avec bras assisté pour la collecte des ordures domestiques* (Study of the biological agents and ergonomic constraints in the use of trucks with articulated arms for collecting household waste). IRSST, Montréal, http://www.irsst.qc.ca/en/_publicationirsst_860.html, Études et recherches / Rapport R-317, 2002

[261] World Health Organisation (WHO), *Managing an injection safety policy*. http://www.who.int/injection_safety/toolbox/en/ManagingInjectionSafety.pdf, WHO/BCT/03.01. 14th Mar 2003

[262] Canadian Centre for Occupational Health and Safety — CCOHS, *OSH answers — AIDS precautions, Needles and sharps*. http://www.ccohs.ca/oshanswers/diseases/aids/health_care2.html#_1_2, Jun 2000

[263] Baur, X., 'Extrinsic allergic alveolitis as an occupational disease', *Zbl Arbeitsmed*. Vol. 46, 1996, pp. 438–42.

[264] Missel, T., Hartung, J., *Partikelzählung zur Erfassung von Schimmelpilzen in der Arbeitsplatzatmosphäre*. Schriftenreihe der Bundesanstalt für Arbeitsschutz und Arbeitsmedizin, Forschung Fb 1043, Dortmund, Berlin, Dresden, http://www.baua.de/nn_28516/de/Informationen-fuer-die-Praxis/Publikationen/Schriftenreihe/Forschungsberichte/2005/Fb1043,xv=vt.pdf, 2005

[265] Steinberg, R., *Keimemissionen in der Abfallwirtschaft unter Berücksichtigung des Arbeitsschutzes*. Veröffentlichung des Fachgebiets Abfall- und Siedlungswissenschaft Bergische Universität, Gesamthochschule Wuppertal, Bd. 2, 1997

[266] Kolk, A., 'Schimmelpilze in Abfallsortieranlagen', Nr. 0075, Ausgabe 1/2006. In: *Aus der Arbeit des BIA*. Hrsg., Berufsgenossenschaftliches Institut für Arbeitsschutz — BGIA, Sankt Augustin — Loseblatt-Ausgabe, Jan 2006

[267] Deininger, C., 'Untersuchungen zur mikrobiellen Luftbelastung in 32 Wertstoffsortieranlagen', *Gefahrstoffe — Reinhalt Luft* Vol. 58, No. 3, 1998, pp. 113–23.

[268] Heida, H. et al., 'Occupational exposure and indoor air quality monitoring in a composting facility', *Am Ind Hyg Assoc J* Vol. 56, No. 1, 1995, pp. 39–43.

[269] Streib, R., Botzenhart, K., Drysch, K., Rettenmeier, A. W., 'Assessment of exposure to dust and microorganisms during delivery, sorting, and composting of domestic and industrial waste materials', *Zentralbl Hyg Umweltmed* Vol. 198, 1996, pp. 531–51.

[270] Wilkins, K., 'Volatile organic compounds from household waste', *Chemosphere* Vol. 29, No. 1, 1994, pp. 47–53.

[271] Tolvanen, O. K. et al., 'Occupational hygiene in biowaste composting', *Waste Management Res* Vol. 16, No. 6, 1998, pp. 525–40.

[272] Komilis, D. P. et al., 'Emission of volatile organic compounds during composting of municipal solid wastes', *Water Res* Vol. 38, 2004, pp. 1707–1714.

[273] Müller, T. et al., '(M) VOC and composting facilities. Part 1: (M) VOC emissions from municipal biowaste and plant refuse', *Environ Pollut Res* Vol. 11, No. 2, 2004, pp. 91–7.

[274] Tolvanen, O. et al., 'Occupational hygiene in a Finnish drum composting plant', *Waste Manag* Vol. 25, 2005, pp. 427–33.

[275] Eitzer, B. D., 'Emissions of volatile organic chemicals from municipal solid waste composting facilities', *Environ Sci Technol* Vol. 29, 1995, pp. 896–902.

[276] Leach, J., et al., Volatile organic compounds in an urban airborne environment adjacent to a municipal incinerator, waste collection centre and sewage treatment plant. *Atmosp Environ* Vol. 33, 1999, pp. 4309–25.

[277] Douwes, J, 't Mannetje, A., Heederik, D., 'Work related symptoms in sewage treatment workers', *Ann Agric Environ Med* Vol. 8, 2001, pp. 39–45, http://www.aaem.pl/pdf/aaem0106.pdf

[278] Jager, E., Eckrich, C., 'Hygienic aspects of biowaste composting', *Ann Agric Environ Med* Vol. 4, 1997, pp. 99–105.

[279] Lavoie, J., Alie, R., 'Determining the characteristics to be considered from a worker health and safety standpoint in household waste sorting and composting plants', *Ann Agric Environ Med* Vol. 4, 1997, pp. 123–8.

[280] van Tongeren, M. et al., 'Exposure to organic dusts, endotoxins and microorganisms in the municipal waste industry', *Int J Occup Environ Health* Vol. 3, No. 1, 1997, pp. 30–36, http://www.ncbi.nlm.nih.gov/entrez/query.fcgi?cmd=Retrieve&db=PubMed&list_uids=9891098&dopt=Abstract

[281] Sigsgaard, T. et al., 'Biomonitoring and work related symptoms among garbage handling workers', *Ann Agric Environ Med* Vol. 4, 1997, pp. 107–12.

[282] Rahkonen, P., 'Airborne contaminants at waste treatment plants', *Waste Management Research* Vol. 10, 1992, pp. 411–21.

[283] Würtz, E., Breum, N. O., 'Exposure to microorganisms during manual sorting of recyclable paper of different quality', *Ann Agric Environ Med* Vol. 4, 1997, pp. 129–35, http://www.aaem.pl/pdf/9741_129.pdf

[284] Millner, P. D. et al., 'Bioaerosols associated with composting facilities', *Compost Science Utilisation* Vol. 2, No. 4, 1994, pp. 6–57.

[285] Bünger, J., Antlauf-Lammers, M., Schulz, T. G., Westphal, G. A., Müller, M. M., Ruhnau, P., Hallier, E., 'Health complaints and immunological markers of exposure to bioaerosols among biowaste collectors and compost workers', *Occup Environ Med* Vol. 57, 2000, pp. 458–64.

[286] Nowak, D., Garz, S., Schottky, A., 'Zur Bedeutung von Endotoxinen für obstruktive Atemwegserkrankungen im Bereich der Landwirtschaft', *Arbeitsmed Sozialmed Umweltmed* Vol. 33, 1998, pp. 233–40.

[287] Wouters, I. M. et al., 'Upper airway inflammation and respiratory symptoms in domestic waste collectors', *Occup Environ Med* Vol. 59, No. 2, 2002, pp. 106–12.

[288] Swan, J. R. M., Kelsey, A., Crook, B., Gilbert, E. J., *Occupational and environmental exposure to bioaerosols from composts and potential health effects — A critical review of published data*. Health and Safety Executive — HSE, Research report 130, 2003, ISBN 0 7176 2707 1, http://www.hse.gov.uk/research/rrpdf/rr130.pdf,

[289] Prasad, M., van der Werf, P., Brinkmann, A., *Bioaerosols and composting — A literature evaluation*. Composting Association of Ireland, http://www.compostireland.ie/docs/cre_bioaerosol_aug2004.pdf, Aug 2004

[290] Sigsgaard, T., Malmros, P., Nersting, L., Petersen, C., 'Respiratory disorders and atopy in Danish refuse workers', *Am J Respir Crit Care Med* Vol. 149, 1994, pp. 1407–12.

[291] Kuchuk, A. A., Basanets, A., Louhelainen, K., 'Bronchopulmonary pathology in workers exposed to organic fodder dust', *Ann Agric Environ Med* Vol. 7, 2000, pp. 17–23.

[292] Baur, X., Schneider, W. D., 'Non-Allergic Obstructive Respiratory Tract Disease in Agriculture', *Pneumologie* Vol. 54, 2000, pp. 80–91.

[293] Fischer, G., Müller, T., Schwalbe, R., Ostrowski, R., Dott, W., 'Exposure to airborne fungi, MVOC and mycotoxins in biowaste-handling facilities', *Int J Hyg Environ Health* Vol. 203, 2000, pp. 97–104.

[294] Rylander, R., Lin, R. H., ' (1?3)-ß-D-glucans — relationship to indoor air-related symptoms, allergy and asthma', *Toxicology* Vol. 152, 2000, pp. 47–52.

[295] Charbotel, B., Hoeurs, M., Perdrix, A., Anzivino-Viricel, L., Bergeret, A., 'Respiratory function among waste incinerator workers', *International Archives of Occupational and Environmental Health* Vol. 78, 2005, pp. 65–70.

[296] Bünger, J., Schappler-Scheele, B., Hilgers, R., Hallier, E., ,Verschlechterung der Lungenfunktion von Kompostwerkern durch Bioaerosole', *Arbeitsmed. Sozialmed. Umweltmed.* Vol. 38, 2003, p. 141.

[297] Bünger, J., Schappler-Scheele, B., Missel, T., Hilgers, R., Kämpfer, S., Felten, C., Leifert, I., Hasenkamp, P., *Health risks in composting plants from biological agents: 5-year-follow-up*. Bremerhaven, Wirtschaftsverlag NW Verlag für neue Wissenschaft GmbH, http://www.baua.de/nn_5846/en/Practical-experience/Publications/Institute-Report-Series/Research-reports/2003/Fb_20993.html__nnn=true, Schriftenreihe der Bundesanstalt für Arbeitsschutz und Arbeitsmedizin: Forschungsbericht, Fb 993, Edition 1, 2003.

[298] Baur, X., Richter, G., Pethran, A., Czuppon, A. B., Schwaiblmair, M., 'Increased prevalence of IgG-induced sensitization and hypersensitivity pneumonitis (humidifier lung) in nonsmokers exposed to aerosols of a contaminated air conditioner', *Respiration* Vol. 59, 1992, pp. 211–14.

[299] Technische Regeln für Biologische Arbeitsstoffe, 'Abfallsortieranlagen — Schutzmaßnahmen' (*TRBA* 210). BArbBl. (1999), 6, 77–81, zuletzt geändert BArbBl., 8, 79, 2001.

[300] Technische Regeln für Biologische Arbeitsstoffe, 'Biologische Abfallbehandlungsanlagen — Schutzmaßnahmen' (*TRBA* 211). BArbBl. (2002), 10, 21 pp, zuletzt geändert BArbBl. 10, 84–87, 2002.

[301] Technische Regeln für Biologische Arbeitsstoffe, 'Thermische Abfallbehandlung — Schutzmaßnahmen' (*TRBA* 212). BArbBl., 10, 39–44, 2003.

[302] Technische Regeln für Biologische Arbeitsstoffe, 'Abfallsammlung — Schutzmaßnahmen' (TRBA 213). BArbBl., 8/9, 53–57, 2005.

[303] Missel, T., Deininger, C., 'Beurteilung der Wirksamkeit keimemissions-mindernder Maßnahmen in Wertstoffsortieranlagen', In: BIA-Handbuch, 29. Lieferung VII/97. Hrsg. Berufsgenossenschaftliches Institut für Arbeitsschutz — BGIA, Sankt Augustin, Erich Schmidt, Bielefeld — Loseblatt-Ausgabe, 1997.

[304] Rosenfeld, P. et al., 'Measurement of biosolids compost odor emissions from a windrow static pile and biofilter', *Water Environ Res* Vol. 76, No. 4, 2004, pp. 310–15.

[305] Lavoie, J., Guertin, S., 'Evaluation of health and safety risks in municipal solid waste recycling plants', *J Air Waste Manag Assoc* Vol. 51 (March 2001), pp. 352–60.

[306] Dorevitsch, S., Marder, D., 'Occupational hazards of municipal solid waste workers', *Occup Med* Vol. 16, 2001, pp. 125–33.

[307] Directive 2000/54/EC of the European Parliament and of the Council of 18 September 2000 on the protection of workers from risks related to exposure to biological agents at work (seventh individual directive within the meaning of Article 16(1) of Directive 89/391/EEC). *Official Journal* L 262, 17/10/2000, pp. 21-45, http://europa.eu.int/smartapi/cgi/sga_doc?smartapi!celexapi!prod!CELEXnumdoc&lg=en&numdoc=32000L0054&model=guichett

[308] Commission Directive 93/67/EEC of the 20th July 1993 laying down the principles for assessment of risks to man and the environment of substances notified in accordance with Council Directive 67/548/EEC. *Official Journal* L 227, 08/09/1993, pp. 9-18, http://europa.eu.int/eur-lex/lex/LexUriServ/LexUriServ.do?uri=CELEX:31993L0067:EN:HTML

[309] European Agency for Safety and Health at Work, 'Dangerous substances — Handle with care', *Magazine* 6, 2003, http://osha.eu.int/publications/magazine/6/magazine6-en.pdf, ISSN 1608-4144

[310] International Labour Office (ILO), *Biological Hazards. Encyclopaedia of Occupational Health and Safety* (4th edn), 1998, http://www.ilo.org/encyclopaedia

[311] AUVA, *Biologische Arbeitsstoffe — Evaluierungsheft E 05*, http://www.auva.at/mediaDB/48843.PDF (language: de), 2005

[312] ISPESL, *Guidelines: risk assessment on biotechnology*, http://www.ispesl.it/linee_guida/fattore_di_rischio/ biotec.htm, 2005

[313] EN 13098, *Workplace atmosphere — Guidelines for measurement of airborne micro-organisms and endotoxin*

[314] EN 14031, *Workplace atmosphere — Measurement of endotoxin*

[315] EN 14042, *Workplace atmospheres. Guide for the application and use of procedures for the assessment of exposure to chemical and biological agents*

[316] Dutkiewicz, J, Jablonski, L, Olenchock, S-A., 'Occupational biohazards. A review', *Am J Ind Med* Vol. 14, 1988, pp. 605–23.

[317] Eduard, W., Heederik, D., Methods for quantitative assessment of airborne levels of non-infectious micro-organisms in highly contaminated work environments. *Am Ind Hyg Assoc J* Vol. 59, 1998, pp. 113-27.

[318] Kolk, A., 'Verfahren zur Bestimmung der Bakterienkonzentration in der Luft am Arbeitsplatz', Kennzahl 9430. In: *BGIA-Arbeitsmappe Messung von Gefahrstoffen*. Hrsg., Berufsgenossenschaftliches Institut für Arbeitsschutz — BGIA

[319] Kämpfer, P., Beyer, W., Danneberg, G., Grün, L., Martens, W., Neef, A., Palmgren, U., Szewzyk, R., 'Neuere Methoden zum Nachweis luftgetragener Mikroorganismen und zur Quellenidentifikation', In: *Stand von Wissenschaft, Forschung und Technik zu siedlungshygienischen Aspekten der Abfallentsorgung und –Verwertung*. Tagung 30. August bis 1. September 1999 (Hrsg. Eikmann, Th. und R. Hofmann). Schriftenreihe des Vereins für Wasser-, Boden- und Lufthygiene. Band 30, 321–402.

[320] Jäckel, U., P. Kämpfer, 'Grenzen und Möglichkeiten zur Detektion luftgetragener Bakterien', In: *KRdL-Experten-Forum Mikrobakterielle Luftverunreinigungen*, 13./14. Oktober 2005, Freising — Weihenstephan. KRdL-Schriftenreihe Band 35, 117–143.

[321] Boleij, J., Buringh, E., Heederik, D., Kromhout, H., *Occupational Hygiene of Chemical and Biological Agents*. Elsevier, Amsterdam, 1995.

[322] Thorne, P. S., Lange, J. L., Bloebaum, P. D., Kullman, G. J., 'Bioaerosol sampling in field studies: can samples be express mailed?' *Am Ind Hyg Assoc J* Vol. 55, 1994, pp. 1072–9.

[323] Lange, J. L., Thorne, P. S., Lynch, N. L., 'Application of flow cytometry and fluorescent in situ hybridization for assessment of exposures to airborne bacteria', *Appl Environ Microbiol* Vol. 63, 1997, pp. 1557–63.

[324] Górny, R. L., Dutkiewicz, J., Krysińska-Traczyk, E., 'Size distribution of bacterial and fungal bioaerosols in indoor air', *Ann Agric Environ Med* Vol. 6, 1998, pp. 105–13.

[325] Reponen, T. T., Willeke, K., Grinshpun, S. A., 'Biological particles sampling', In: Baron, P. A., Willeke, K. (eds), *Aerosol Measurement: Principles, Techniques and Applications*, Wiley-Interscience, NY, 2001, pp. 751–78.

[326] Duchaine, C., Thorne, P. S., Meriaux, A., Grimard, Y., Whitten, P., Cormier, Y., 'Comparison of endotoxin exposure assessment by bioaerosol impinger and filter sampling methods', *Appl Environ Microbiol* Vol. 67, 2001, pp. 2775–80.

[327] Mainelis, G., Górny, R., Willeke, K., Reponen, T., 'Rapid counting of liquid-borne microorganisms by light scattering spectrometry', *Ann Agric Environ Med* Vol. 12, No. 1, 2005, pp. 141–8.

[328] Eduard, W., Sandven, P., Johansen, B. V., Bruun, R., 'Identification and quantification of mould spores by scanning electron microscopy (SEM): analysis of filter samples collected in Norwegian saw mills', *Ann Occup Hyg* Vol. 31, 1988, pp. 447–55.

[329] Karlsson, K., Malmberg, P., 'Characterization of exposure to molds and actinomycetes in agricultural dusts by scanning electron microscopy, fluorescence microscopy and the culture method', *Scand J Work Environ Health* Vol. 15, 1989, pp. 353–9.

[330] Henningson, E. W., Lundquist, M., Larsson, E., Sandstrom, G., Forsman, M., 'A comparative study of different methods to determine the total number and the survival ratio of bacteria in aerobiological samplers', *J Aerosol Sci* Vol. 28, 1997, pp. 459–69.

[331] Dillon, H. K., Heinsohn, P. A., Miller, J. D., *Field guide for the determination of biological contaminants in environmental samples*. American Industrial Hygiene Association, Fairfax, VA, 1996.

[332] Miller, J. D., Young, J. C., 'The use of ergosterol to measure exposure to fungal propagules in indoor air', *Am Ind Hyg Assoc J* Vol. 58, 1997, pp. 39–43.

[333] Douwes, J., van der Sluis, B., Doekes, G., 'Fungal extracellular polysaccharides in house dust as a marker for exposure to fungi: relations with culturable fungi, reported home dampness and respiratory symptoms', *J Allergy Clin Immunol* Vol. 103, 1999, pp. 494–500.

[334] Khan, A. A., Cerniglia, C. E., 'Detection of Pseudomonas aeruginosa from clinical and environmental samples by amplification of the exotoxin A gene using PCR', *Appl Environ Microbiol* Vol. 60, 1994, pp. 3739–45.

[335] Alvarez, A. J., Buttner, M. P., Toranzos, G. A., 'Use of solid-phase PCR for enhanced detection of airborne micro-organisms', *Appl Environ Microbiol* Vol. 60, 1994, pp. 374–6.

[336] Bang, F. B., 'A bacterial disease of Limulus polyphemus', *Bull Johns Hopkins Hosp*, Vol. 98, 1956, pp. 325–50.

[337] Sonesson, A., Larsson, L., Fox, A., Westerdahl, G., Odham, G., 'Determination of environmental levels of peptidoglycan and lipopolysaccharide using gas chromatography-mass spectrometry utilizing bacterial amino acids and hydroxy fatty acids as biomarkers', *J Chromatogr Biomed Appl* Vol. 431, 1988, pp. 1–15.

[338] Sonesson, A., Larsson, L., Schütz, A., Hagmar, L., Hallberg, T., 'Comparison of the limulus amebocyte lysate test and gas chromatography-mass spectrometry for measuring lipopolysaccharides (endotoxins) in airborne dust from poultry-processing industries', *Appl Environ Microbiol* Vol. 56, 1990, pp. 1271–8.

[339] Aketagawa, J., Tanaka, S., Tamura, H., Shibata, Y., Sait, H., 'Activation of limulus coagulation factor G by several (1?3)-ß-D-glucans: comparison of the potency of glucans with identical degree of polymerization but different conformations', *J Biochem* Vol. 113, 1993, pp. 683–6.

[340] Douwes, J., Doekes, G., Montijn, R., Heederik, D., Brunekreef, B., 'Measurement of ß(1?3)-glucans in the occupational and home environment with an inhibition enzyme immunoassay', *Appl Environ Microbiol* Vol. 62, 1996, pp. 3176–82.

[341] Burell, R., Rylander, R.,'A critical review of the role of precipitins in hypersensitivity pneumonitis', *Eur Respir Dis* Vol. 62, 1981, pp. 332–43.

[342] Eduard, W., Sandven, P., Levy, F., 'Relationships between exposure to spores from Rhizopus microsporus and Paecilomycetes variotti and serum IgG antibodies in wood trimmers', *Int Arch Allergy Immunol* Vol. 97, 1992, pp. 274–82.

[343] Luczynska, C. M., Arruda, L. K., Platts-Mills, T. A., et al., 'A two-site monoclonal antibody ELISA for the quantification of the major Dermatophagoides spp. allergens, Der p I and Der f I', *J Immunol Methods* Vol. 118, 1989, pp. 227–35.

[344] Price, J. A., Pollock, I., Little, S. A., Longbottom, J. L., Warner, J. O., 'Measurement of airborne mite antigen in homes of asthmatic children', *Lancet* Vol. 336, 1990, pp. 895–7.

[345] Houba, R., van Run, P., Doekes, G., Heederik, D., Spithoven, J., 'Airborne levels of alpha-amylase allergens in bakeries', *J Allergy Clin Immunol* Vol. 99, 1997, pp. 286–92.

[346] Leaderer, B. P., Belanger, K., Triche, E., et al., 'Dust mite, cockroach, cat, and dog allergen concentrations in homes of asthmatic children in the northeastern United States: impact of socioeconomic factors and population density', *Environ Health Perspect* Vol. 110, 2002, pp. 419–25.

[347] Swanson, M. C., Agarwal, M. K., Reed, C. E., 'An immunochemical approach to indoor aeroallergen quantitation with a new volumetric air sampler: studies with mice, roach, cat, mouse, and guinea pig antigens', *J Allergy Clin Immunol* Vol. 76, 1985, pp. 724–9.

[348] Virtanen, T., Louhelainen, K., Montyjarvi, R., 'Enzyme-linked immunosorbent assay (ELISA) inhibition method to estimate the level of airborne bovine epidermal antigen in cowsheds', *Int Arch Allergy Appl Immunol* Vol. 81, 1986, pp. 253–7.

[349] Schou, C., Svendsen, U. G., Lowenstein, H., 'Purification and characterization of the major dog allergen, Can f I', *Clin Exp Immunol* Vol. 21, 1991, pp. 321–8.

[350] Pollart, S. M., Smith, T. F., Morris, E. C., Gelber, L. E., Platts-Mills, T. A., Chapman, M. D., 'Environmental exposure to cockroach allergens: analysis with monoclonal antibody-based enzyme immunoassays', *J Allergy Clin Immunol* Vol. 87, 1991, pp. 505–10.

[351] Miguel, A. G., Cass, G. R., Weiss, J., Glovsky, M. M., 'Latex allergens in tire dust and airborne particles', *Environ Health Perspect* Vol. 104, 1996, pp. 1180–86.

[352] Thorne, P. S., Reynolds, S. J., Milton, D. K., et al., 'Field evaluation of endotoxin air sampling assay methods', *Am Ind Hyg Assoc J* Vol. 58, 1997, pp. 792–9.

[353] Chun, D. T., Chew, V., Bartlett, K., et al., 'Preliminary report on the results of the second phase of a round-robin endotoxin assay study using cotton dust', *Appl Occup Environ Hyg* Vol. 15, 2000, pp. 152–7.

[354] Reynolds, S., Thorne, P., Donham, K., et al., 'Interlaboratory comparison of endotoxin assays using agricultural dusts', *Am Ind Hyg Assoc J* Vol. 63, 2002, pp. 430–38.

[355] Douwes, J., Versloot, P., Hollander, A., Heederik, D., Doekes, G., 'Influence of various dust sampling and extraction methods on the measurement of airborne endotoxin', *Appl Environ Microbiol* Vol. 61, 1995, pp. 1763–9.

[356] Iversen, M., Pedersen, B., 'The prevalence of allergy in Danish farmers', *Allergy* Vol. 45, 1999, pp. 347–53.

[357] Hollander, A., Heederik, D., Doekes, G., 'Respiratory allergy to rats: exposure–response relationships in laboratory animal workers', *Am J Respir Crit Care Med* Vol. 155, 1997, pp. 562–7.

[358] Mo3ocznik, A., 'Time of farmers' exposure to biological factors in agricultural working environment', *Ann Agric Environ Med* Vol. 11, No. 1, 2004, pp. 85–9, http://www.aaem.pl/pdf/11085.htm

[359] Górny, R. L., 'Biohazards: standards, guidelines, and proposals for threshold limit values', *Principles and methods of assessing the working environment* Vol. 41, No. 3, 2004, pp. 17–40

[360] Górny, R. L., Dutkiewicz, J., 'Bacterial and fungal aerosol in indoor environment in Central and Eastern European countries', *Ann Agric Environ Med* Vol. 9, 2002, pp. 17–23.

[361] Sociaal Economische Raad (SER), *MAC-Waarden, Aflatoxinen*, http://www.ser.nl/overdeser/default.asp?desc=mac_waarden_aflatoxinen

European Agency for Safety and Health at Work

Expert forecast on Emerging Biological Risks related to Occupational Safety and Health

Luxembourg: Office for Official Publications of the European Communities

2007 — 145 pp. — 21 x 29.7 cm

ISBN 92-9191-130-5

Price (excluding VAT) in Luxembourg: EUR 15